Praise for David Reich's

Who We Are and How We Got Here

"In *Who We Are and How We Got Here: Ancient DNA and the New Science of the Human Past*, David Reich . . . introduces us to the twenty-first century Rosetta Stone: ancient DNA, which will do more for our understanding of prehistory than radiocarbon dating did. . . . *Who We Are and How We Got Here* is less than 300 pages of text, but it is packed with startling facts and novel revelations that overturn the conventional expectations of both science and common sense."

—*National Review*

"Remarkable. . . . Spectacular. . . . In making constant new discoveries about humanity, Reich and his Harvard team are now plunging into uncharted academic waters. . . . Reich's influence in this field has been immense and the output of his department monumental. . . . Thrilling in its clarity and its scope."

—*The Guardian*

"This is a compendious book . . . its importance cannot be overstated and neither can some of its best stories."

—*The Sunday Times* (London)

"[A] thrilling account of mapping humans through time and place. . . . Genomics and statistics have drawn back the curtain on the sort of sex and power struggles you'd expect in *Game of Thrones*. . . . We do need a non-loaded way to talk about genetic diversity and similarities in populations. This book goes some way to starting that conversation."

—*Nature*

"In this comprehensive and provocative book, David Reich exhumes and examines fundamental questions about our origin and future using powerful evidence from human genetics. What does 'race' mean in 2018? How alike and how unlike are we? What does identity mean? Reich's book is sobering and clear-eyed, and, in equal parts, thrilling and thought-provoking. There were times that I had to stand up and clear my thoughts to continue reading this astonishing and important book."   —Siddhartha Mukherjee, Pulitzer Prize–winning
author of *The Emperor of All Maladies*

"Reich's book reads like notes from the frontline of the 'Ancient DNA Revolution' with all the spellbinding drama and intrigue that come with such a huge transformation in our understanding of human history."   —Anne Wojcicki, CEO and cofounder of 23andMe

"This absorbing book will blow you away with its rich and astounding account of where we came from and why that matters. Reich tells the surprising story of how humans got to every corner of the planet, which was revealed only after he and other scientists unlocked the secrets of ancient DNA. The courageous, compassionate, and highly personal climax will transform how you think about the meaning of ancestry and race."
—Daniel E. Lieberman, professor of human evolutionary biology,
Harvard University, and author of *The Story of the
Human Body: Evolution, Health, and Disease*

David Reich

Who We Are and How We Got Here

David Reich, professor of genetics at Harvard Medical School and a Howard Hughes Medical Institute Investigator, is one of the world's leading pioneers in analyzing ancient human DNA. In a 2015 article in *Nature*, he was named one of "ten people who matter" in all of the sciences for his contribution to transforming ancient DNA data "from niche pursuit to industrial process." Awards he has received include the Newcomb Cleveland Prize from the American Association for the Advancement of Science and the Dan David Prize in the Archaeological and Natural Sciences for his computational discovery of intermixing between Neanderthals and *Homo sapiens*.

Who We Are
and
How We Got Here

*Ancient DNA and the New Science
of the Human Past*

David Reich

Vintage Books
A Division of Penguin Random House LLC
New York

The Library of Congress has cataloged the Pantheon edition as follows:
Name: Reich, David (of Harvard Medical School), author.
Title: Who we are and how we got here : ancient DNA revolution and the
new science of the human past / David Reich.
Description: First edition. New York : Pantheon Books, 2018. Includes
bibliographical references and index.
Identifiers: LCCN 2017038165
Subjects: LCSH: Human genetics—Popular works. Genomics—Popular
works. DNA—Analysis. Prehistoric peoples. Human population genetics.
BISAC: SCIENCE / Life Sciences / Genetics & Genomics. SCIENCE / Life
Sciences / Evolution. SOCIAL SCIENCE / Anthropology / General.
Classification: LCC QH431 .R37 2018. | DDC 572.8/6—dc23.
LC record available at lccn.loc.gov/2017038165

Vintage Books Trade Paperback ISBN: 978-1-101-87346-5
eBook ISBN: 978-1-101-87033-4

Author photo © Bizu Tesfaye/Howard Hughes Medical Institute
Illustrations and map by Oliver Uberti

www.vintagebooks.com

Printed in the United States of America
10

For Seth and Leah

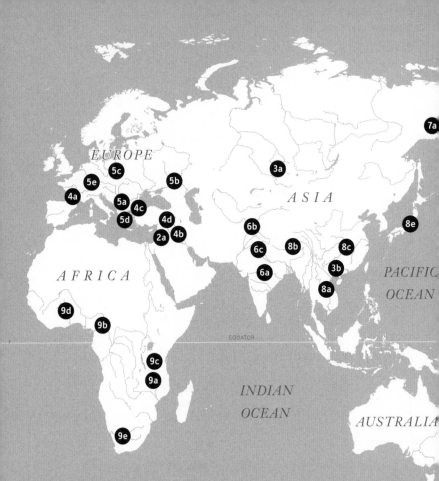

30 Population Mixtures

The mixture of highly differentiated populations is a recurrent process in our history. This map provides a key to thirty great mixture events discussed in this book. (Locations are not meant to be precise.)

CHAPTER 2

2a 54,000–49,000 years ago
All non-Africans
Neanderthals + modern humans

CHAPTER 3

3a >70,000 ya
Siberian Denisovans
*Superarchaic lineage +
Neanderthal-related lineage*

3b 49,000–44,000 ya
Papuans and Australians
Denisovans + modern humans

ARCTIC OCEAN

NORTH AMERICA

ATLANTIC OCEAN

SOUTH AMERICA

CHAPTER 4

4a 19,000–14,000 ya
Magdalenian expansion
Aurignacian + Gravettian lineages

4b >14,000 ya
Late Near Eastern hunter-gatherers
Basal Eurasians + early Near Eastern hunter-gatherers

4c ~14,000 ya
Bølling-Allerød expansion
Southwest + Southeast European hunter-gatherers

4d 8,000–3,000 ya
Copper and Bronze Age Near East
Iranian + Levantine + Anatolian farmers

CHAPTER 5

5a 9,000–5,000 ya
First European farmers
Local hunter-gatherers + Anatolian farmers

5b 9,000–5,000 ya
Steppe pastoralists
Iranian farmers + local hunter-gatherers

5c 5,000–4,000 ya
Northern European Bronze Age
Eastern European farmers + steppe pastoralists

5d >3,500 ya
Aegean Bronze Age
Iranian farmers + European farmers

5e 3,500 ya – present
Present-day Europeans
Northern + Southern European Bronze Age populations

CHAPTER 6

6a >4,000 ya
Ancestral South Indians
Iranian farmers + indigenous Indian hunter-gatherers

6b 4,000–3,000 ya
Ancestral North Indians
Steppe pastoralists + Iranian farmers

6c 4,000–2,000 ya
Present-day Indians
Ancestral South Indians + Ancestral North Indians

CHAPTER 7

7a >15,000 ya
First Americans
Ancient North Eurasians + East Asians

7b 5,000–4,000 ya
Paleo-Eskimos
Far Eastern Siberians + First Americans

7c >4,000 ya
Amazonians
Population Y + First Americans

7d 2,000–1,000 ya
Na-Dene speakers
Paleo-Eskimos + First Americans

7e 2,000–1,000 ya
Neo-Eskimos
Far Eastern Siberians + First Americans

CHAPTER 8

8a 5,000–4,000 ya
Austroasiatic speakers
Yangtze River Ghost Population + indigenous Southeast Asian hunter-gatherers

8b 5,000–3,000 ya
Tibetans
Yellow River Ghost Population + Tibetan hunter-gatherers

8c 5,000–1,000 ya
Present-day Han Chinese
Yellow + Yangtze River Ghost Populations

8d 4,000–1,000 ya
Southwest Pacific islanders
Papuans + East Asians

8e 3,000–2,000 ya
Present-day Japanese
Mainland farmers + local hunter-gatherers

CHAPTER 9

9a >8,000 ya
Malawi hunter-gatherers
East + South African foragers

9b 4,000–1,000 ya
Bantu expansion
Cameroon source population + local groups throughout eastern and southern Africa

9c >3,000 ya
East African pastoralists
Levantine farmers + East African foragers

9d >2,000 ya
Present-day West Africans
At least two ancient African lineages

9e 2,000–1,000 ya
Present-day Khoe-Kwadi herders
East African pastoralists + indigenous San

Contents

Acknowledgments

First thing first. This book emerged out of a year of intense collaboration with my wife, Eugenie Reich. We researched the book together, prepared the first drafts of the chapters together, and talked about the book incessantly as it matured. This book would not have come into being without her.

I am grateful to Bridget Alex, Peter Bellwood, Samuel Fenton-Whittet, Henry Louis Gates Jr., Yonatan Grad, Iosif Lazaridis, Daniel Lieberman, Shop Mallick, Erroll McDonald, Latha Menon, Nick Patterson, Molly Przeworski, Juliet Samuel, Clifford Tabin, Daniel Reich, Tova Reich, Walter Reich, Robert Weinberg, and Matthew Spriggs for close critical readings of the entire book.

I thank David Anthony, Ofer Bar-Yosef, Caroline Bearsted, Deborah Bolnick, Dorcas Brown, Katherine Brunson, Qiaomei Fu, David Goldstein, Alexander Kim, Carles Lalueza-Fox, Iain Mathieson, Eric Lander, Mark Lipson, Scott MacEachern, Richard Meadow, David Meltzer, Priya Moorjani, John Novembre, Svante Pääbo, Pier Palamara, Eleftheria Palkopoulou, Mary Prendergast, Rebecca Reich, Colin Renfrew, Nadin Rohland, Daniel Rozas, Pontus Skoglund, Chuanchao Wang, and Michael Witzel for critiques of individual chapters. I also thank Stanley Ambrose, Graham Coop, Dorian Fuller, Éadaoin Harney, Linda Heywood, Yousuke Kaifu, Kristian Kristiansen, Michelle Lee, Daniel Lieberman, Michael McCormick, Michael Petraglia, Joseph Pickrell, Stephen Schiffels, Beth Shapiro, and Bence Viola for reviewing sections of the book for accuracy.

I am grateful to Harvard Medical School, the Howard Hughes Medical Institute, and the National Science Foundation, all of which generously supported my science while I was working on this project, and viewed it as complementary to my primary research.

I finally thank several people who repeatedly encouraged me to write this book. I resisted the idea for years because I did not want to distract myself from my science, and because for geneticists papers are the currency, not books. But my mind changed as my colleagues grew to include archaeologists, anthropologists, historians, linguists, and others eager to come to grips with the ancient DNA revolution. There are many papers I did not write, and many analyses I did not complete, because of the time I needed to write this book. I hope that those who read the book will emerge with a new perspective on who we are.

Introduction

This book is inspired by a visionary, Luca Cavalli-Sforza, the founder of genetic studies of our past. I was trained by one of his students, and so it is that I am part of his school, inspired by his vision of the genome as a prism for understanding the history of our species.

The high-water mark of Cavalli-Sforza's career came in 1994 when he published *The History and Geography of Human Genes*, which synthesized what was then known from archaeology, linguistics, history, and genetics to tell a grand story about how the world's peoples got to be the way they are today.[1] The book offered an overview of the deep past. But it was based on what was known at the time and was therefore handicapped by the paucity of genetic data then available, which were so limited as to be nearly useless compared to the far more extensive information from archaeology and linguistics. The genetic data of the time could sometimes reveal patterns consistent with what was already known, but the information they provided was not rich enough to demonstrate anything truly new. In fact, the few major new claims that Cavalli-Sforza did make have essentially all been proven wrong. Two decades ago, everyone, from Cavalli-Sforza to beginning graduate students such as myself, was working in the dark ages of DNA.

Cavalli-Sforza made a grand bet in 1960 that would drive his entire career. He bet that it would be possible to reconstruct the great migrations of the past based entirely on the genetic differences among present-day peoples.[2]

Through study after study over the subsequent five decades, Cavalli-Sforza seemed to be well on the path to making good on his bet. When he started his work, the technology for studying human variation was so poor that the only possibility was to measure proteins in the blood, using variations like the A, B, and O blood types that are tested by physicians to match blood donors to recipients. By the 1990s, he and his colleagues had assembled data from more than one hundred such variations in diverse populations. Using these data they were able to reliably cluster individuals by continent based on how often they matched each other at these variations: for example, Europeans have a high rate of matching to other Europeans, East Asians to East Asians, and Africans to Africans. In the 1990s and 2000s, they brought their work to a new level by moving beyond protein variation and directly examining DNA, our genetic code. They analyzed a total of about one thousand individuals from around fifty populations spread across the planet, examining variation at more than three hundred positions in the genome.[3] When they told their computer—which had no knowledge of the population labels—to cluster the individuals into five groups, the results corresponded uncannily well to commonly held intuitions about the deep ancestral divisions among humans (West Eurasians, East Asians, Native Americans, New Guineans, and Africans).

Cavalli-Sforza was especially interested in interpreting the genetic clusters among present-day people in terms of population history. He and his colleagues analyzed their blood group data by using a technique that identifies combinations of biological variations that are most efficient at summarizing differences across individuals. Plotting these combinations of blood group types onto a map of West Eurasia, they found that the one summarizing the most variation across individuals reached its extreme value in the Near East, and declined along a southeast-to-northwest gradient into Europe.[4] They interpreted this as a genetic footprint of the migration of farmers into Europe from the Near East, known from archaeology to have occurred after nine thousand years ago. The declining intensity suggested to them that after arriving in Europe, the first farmers mixed with local hunter-gatherers, accumulating more hunter-gatherer ancestry as they expanded, a process they called "demic diffusion."[5] Until recently, many archaeologists viewed the demic diffusion model as an exemplary merging of insights from archaeology and genetics.

The model that Cavalli-Sforza and colleagues proposed to describe the data was intellectually attractive, but it was wrong. Its flaws became apparent beginning in 2008, when John Novembre and colleagues demonstrated that gradients like those observed in Europe can arise without migration.[6] They then showed that a Near Eastern farming expansion into Europe might counter-intuitively cause the mathematical technique that Cavalli-Sforza used to produce a gradient perpendicular to the direction of migration, not parallel to it as had been seen in the real data.[7]

It took the revolution wrought by the ability to extract DNA from ancient bones—the "ancient DNA revolution"—to drive a nail into the coffin of the demic diffusion model. The ancient DNA revolution documented that the first farmers even in the most remote reaches of Europe—Britain, Scandinavia, and Iberia—had very little hunter-gatherer-related ancestry. In fact, they had less hunter-gatherer ancestry than is present in diverse European populations today. The highest proportion of early farmer ancestry in Europe is today not in Southeast Europe, the place where Cavalli-Sforza thought it was most common based on the blood group data, but instead is in the Mediterranean island of Sardinia to the west of Italy.[8]

The example of Cavalli-Sforza's maps shows why his Sforza's grand bet went sour. He was correct in his assumption that the present-day genetic structure of populations echoes some of the great events in the human past. For example, the lower genetic diversity of non-Africans compared to Africans reflects the reduced diversity of the modern human population that expanded out of Africa and the Near East after around fifty thousand years ago. But the present-day structure of human populations cannot recover the fine details of ancient events. The problem is not just that people have mixed with their neighbors, blurring the genetic signatures of past events. It is actually far more difficult, in that we now know, from ancient DNA, that the people who live in a particular place today almost never exclusively descend from the people who lived in the same place far in the past.[9] Under these circumstances, the power of any study that attempts to reconstruct past population movements from present-

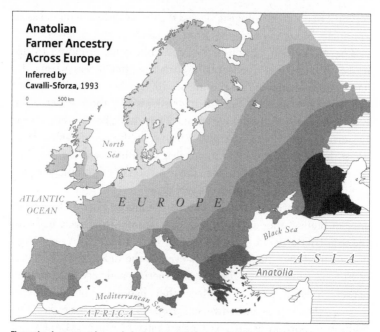

Figure 1a. A contour plot made by Luca Cavalli-Sforza in 1993 (adapted above) suggested that the movement of farmers from the east could be reconstructed from the patterns of blood group variation among people living today, with the highest proportions of such ancestry in the southeast near Anatolia.

day populations is limited. In *The History and Geography of Human Genes*, Cavalli-Sforza wrote that he was excluding from his analysis populations known to be the product of major migrations, such as those of European and African ancestry in the Americas that owe their origin to transatlantic migrations in the last five hundred years, or European minorities such as Roma and Jews. His bet was that the past was a much simpler place than the present, and that by focusing on populations today that are not affected by major migrations in their recorded history, he might be studying direct descendants of people who lived in the same places long before. But what the study of ancient DNA has now shown is that the past was no less complicated than the present. Human populations have repeatedly turned over.

Cavalli-Sforza's transformative contribution to the field of genetic

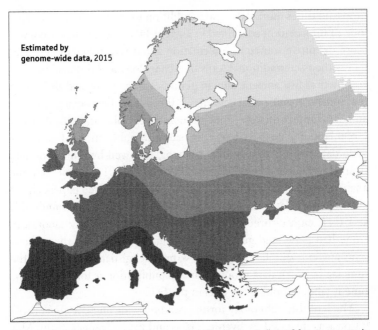

Figure 1b. Modern genome-wide data shows that the primary gradient of farmer ancestry in Europe does not flow southeast-to-northwest but instead in an almost perpendicular direction, a result of a major migration of pastoralists from the east that displaced much of the ancestry of the first farmers.

studies of human prehistory recalls the story of Moses, a visionary leader whose achievement was greater than that of anyone who followed him and who created a new template for seeing the world. The Bible says, "No prophet ever arose again in Israel like Moses," but also tells how Moses was not allowed to reach the promised land. After leading his people for forty years through the wilderness, Moses climbed the mountain of Nebo and looked west over the Jordan River to see the land his people had been promised. But he was not allowed to enter that land. That privilege had been reserved for his successors.

So it is with genetic studies of the past. Cavalli-Sforza saw before anyone else the full potential of genetics for revealing the human past, but his vision predated the technology needed to fulfill it. Today, however, things are very different. We have several hundred thousand times more data, and in addition we have access to the rich lode of information contained in ancient DNA, which has become a more definitive source of information about past population movements than the traditional tools of archaeology and linguistics.

The first five ancient human genomes were published in 2010: a few archaic Neanderthal genomes,[10] the archaic Denisova genome,[11] and an approximately four-thousand-year-old individual from Greenland.[12] The next few years saw the publication of genome-wide data from five additional humans, followed by a burst of data from thirty-eight individuals in 2014. But in 2015, whole-genome analysis of ancient DNA went into hyperdrive. Three papers added genome-wide datasets from another sixty-six,[13] then one hundred,[14] and then eighty-three samples.[15] By August 2017, my laboratory alone had generated genome-wide data for more than three thousand ancient samples. We are now producing data so fast that the time lag between data production and publication is longer than the time it takes to double the data in the field.

Much of the technology for the genome-wide ancient DNA revolution was invented by Svante Pääbo and his colleagues at the Max Planck Institute for Evolutionary Anthropology in Leipzig, Germany, who developed it to study extremely old samples such as archaic Neanderthals and Denisovans. My contribution has been to scale up the methods to study large numbers of relatively more recent samples, albeit still many thousands of years old. The traditional length of an apprenticeship is seven years, and I began mine in 2007 when I started working with Pääbo on the Neanderthal and Denisova genome projects. In 2013, Pääbo helped me to establish my own ancient DNA laboratory—the first in the United States focused on studying whole-genome ancient human DNA. My partner in this effort has been Nadin Rohland, who did her own seven-year

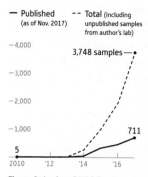

Figure 2. Ancient DNA labs are now producing data so fast that the time lag between data production and publication is longer than the time it takes to double the data in the field.

apprenticeship in Pääbo's laboratory before she came to mine. Our idea was to make ancient DNA industrial—to build an American-style genomics factory out of the techniques developed in Europe to study individual samples.

Rohland and I realized that a technique developed by Matthias Meyer and Qiaomei Fu in Pääbo's laboratory could be the key to the industrial-scale study of ancient DNA. Meyer and Fu's invention was born of necessity: the need to extract DNA from an approximately forty-thousand-year-old early modern human from Tianyuan Cave in China.[16] When Meyer and Fu extracted DNA from Tianyuan's leg bones, they found that only about 0.02 percent of it was from the man himself. The rest came from microbes that had colonized his bones after he died. This made direct sequencing too expensive, even using the hundred-thousand-times cheaper technology that had become available after around 2006. To get around this challenge, Meyer and Fu borrowed a page from the playbook of methods developed by medical geneticists. Just as medical geneticists had developed methods to isolate DNA from the 2 percent of the genome that is most interesting and to discard the other 98 percent, Meyer and Fu isolated a tiny fraction of sequences from the Tianyuan bone that were human and discarded the rest.

The method of DNA isolation that Meyer and Fu developed has been central to the success of the ancient DNA revolution. In the 1990s, molecular biologists learned how to adapt laser-etching techniques invented for printing electronic circuits to attach millions of DNA sequences of their choice to silicon or glass wafers. These sequences could then be cut off the wafers using molecular scissors (enzymes) and released into a watery mix. Meyer and Fu took advantage of this method to synthesize fifty-two-letter-long sequences of DNA that, overlapping like shingles on a roof, covered much of human chromosome 21. Exploiting DNA's tendency to bind to highly similar sequences, they "fished" out the DNA sequences from Tianyuan that they were interested in by using as "bait" the sequences they had artificially synthesized. They found that a large fraction of the DNA they obtained was from Tianyuan's genome. Not only that, but it was from the parts of Tianyuan's genome that they wanted to study. They analyzed the data to show that Tianyuan was an early modern human, part of the lineage leading to present-day East Asians. He did not have a particularly large amount of ancestry from archaic human lineages that were diverged by hundreds of thousands of years from modern human lineages, contradicting earlier claims based on the shape of his skeleton.[17]

Rohland and I adapted this technique to study the whole genome. We worked with our colleagues in Germany to synthesize fifty-two-letter-long DNA sequences covering more than a million positions at which people are known to vary. We used these bait sequences to enrich for human compared to microbial DNA, which in some cases increased the fraction of DNA that was of interest to us by more than a hundredfold. We gained another approximately tenfold jump in efficiency because we only targeted informative positions in the genome. We automated the whole approach, processing the DNA using robots that allowed a single person to study more than ninety samples at once in the span of a few days. We hired a team of technicians to grind powder out of ancient remains, to extract DNA from the powder, and then to turn the extracted DNA into a form that we could sequence. The laboratory work was only the beginning. An equally intricate task was sorting the billions of DNA sequences into the individuals to whom they belonged, analyzing the data and weeding out samples with evidence of contamination, and creating an easily accessible dataset. Shop Mallick, a physicist who had joined my laboratory six years before, set up our computers to do all of this, and continually updated our strategy for processing the data as the nature of the data evolved and its volume increased.

The results were even better than we had hoped. The cost of producing genome-wide data dropped to less than five hundred dollars per sample. This was many dozens of times cheaper than brute-force whole-genome sequencing. Even better, our method made it possible to get genome-wide data out of around half of the skeletal samples we screened, although the success rate of course varied depending on the degree to which the skeletons we examined had been preserved. For example, we have obtained about 75 percent success rates for ancient samples from the cold climate of Russia, but only around 30 percent for samples from the hot Near East.

These advances mean that whole-genome study of ancient DNA

no longer requires screening large numbers of skeletal remains before it is possible to find a few individuals whose DNA can be analyzed. Instead, a substantial fraction of screened samples dating to the last ten thousand years can now be converted to working genome-wide data. The new methods have made it possible to analyze hundreds of samples in a single study. With such data, it is possible to reconstruct population changes in exquisite detail, transforming our understanding of the past.

By the end of 2015, my ancient DNA laboratory at Harvard had published more than half of the world's genome-wide human ancient DNA. We discovered that the population of northern Europe was largely replaced by a mass migration from the eastern European steppe after five thousand years ago[18]; that farming developed in the Near East more than ten thousand years ago among multiple highly differentiated human populations that then expanded in all directions and mixed with each other along with the spread of agriculture[19]; and that the first human migrants into the remote Pacific islands beginning around three thousand years ago were not the sole ancestors of the present-day inhabitants.[20] In parallel, I initiated a project to survey the diversity of the world's present-day populations, using a microchip for analyzing human variation that my collaborators and I designed specifically for the purpose of studying the human past. We used the chip to study more than ten thousand individuals from more than a thousand populations worldwide—a dataset that has become a mainstay of studies of human variation not just in my laboratory but also in other laboratories around the world.[21]

The resolution with which this revolution has allowed us to reconstruct events in the human past is stunning. I remember a dinner at the end of graduate school with my Ph.D. supervisor, David Goldstein, and his wife, Kavita Nayar, both of whom had been students of Cavalli-Sforza. It was 1999, a decade before the advent of genome-wide ancient DNA, and we daydreamed together, wondering how accurately events of the past could be reconstructed by traces left behind. After a grenade explosion in a room, could the exact position of each object prior to the explosion be reconstructed by piecing together the scattered remains and studying the shrapnel in the wall? Could languages long extinct be recalled by unsealing a cave still reverberating with the echoes of words spoken there thousands of years ago? Today, ancient DNA is enabling this kind of detailed reconstruction of deep relationships among ancient human populations.

These days, human genome variation has surpassed the traditional toolkit of archaeology—the study of the artifacts left behind by past societies—in what it can reveal of changes in human populations in the deep past.[22] This has come as a surprise to nearly everyone. Carl Zimmer, a science journalist at *The New York Times* who has written frequently about this new field, told me that when he was assigned by his newspaper to cover the study of ancient DNA, he agreed to do it as a service to the science team, thinking it would be a minor sideshow to his main focus on evolution and human physiology. He imagined writing an article about the field every six months or so, and that the rush of discoveries would end after a year or two. Instead, Zimmer now finds himself dealing with a major new scientific paper every few weeks, even as developments are accelerating and the revolution intensifies.

This book is about the genome revolution in the study of the human past. This revolution consists of the avalanche of discoveries based on data taken from the whole genome—meaning, the entire genome analyzed at once instead of just small stretches of it such as mitochondrial DNA. The revolution has been made far more powerful by the new technologies for extracting whole genomes' worth of DNA from ancient humans. I make no attempt to trace the history of the field of genetic studies of the past—the decades of scientific analysis of human variation that began with studies of skeletal variation and continued with studies of genetic variation in tiny snippets of the human genome. These efforts provided insights into population relationships and migrations, but those insights pale when compared to the dazzling information provided by the extraordinary tranches of data that began to be available after 2009. Before and after that year, studies of one or a few locations in the genome were occasionally the basis for important discoveries, providing evidence in favor of some scenarios over others. Yet genetic evidence before around 2009 was mostly incidental to studies of the human past in other fields, a poor handmaiden to the main business of archaeology. Since 2009, though, whole-genome data have begun to challenge long-held views in archaeology, history, anthropology, and even linguistics—and to resolve controversies in those fields.

The ancient DNA revolution is rapidly disrupting our assumptions about the past. Yet there is at present no book by a working geneticist that lays out the impact of the new science and explains how it can be used to establish compelling new facts. The findings needed to grasp the scope of the ancient DNA revolution are scattered among hard-to-read, jargon-filled scientific papers, sometimes supplemented by hundreds of pages of dense notes on methodology. In *Who We Are and How We Got Here*, I aim to offer readers a clear view through this extraordinary window into the past—to provide a book about the ancient DNA revolution intended for lay reader and specialist alike. My goal is not to present a synthesis—the field is moving too quickly. By the time this book reaches readers, some advances that it describes will have been superseded or even contradicted. In the three years since I began writing, many fresh findings have emerged, so that most of what I describe here is based on results obtained after I started. I hope that readers will take the topics I discuss as examples of the disruptive power of whole-genome studies, not as a definitive summary of the state of the science.

My approach is to take readers through the process of discovery, with each chapter serving as an argument that has as its goal to bring readers, who may have come with one perspective when they started, to another place when they finish. I try to make a virtue of my laboratory's central role in the ancient DNA revolution by telling the story of my own work where it is relevant—as this is a subject on which I can speak with great authority—while also discussing work in which I was not involved when it is critical to the story. Because I take this approach, the book disproportionately highlights the work from my laboratory. I apologize that I have been able to mention by name only a tiny fraction of the people who made equally important contributions. My priority has been to convey the excitement and surprise of the genome revolution, and to take readers on a compelling narrative path through it, not to write a scientific review.

I also highlight some of the great themes that are emerging, espe-

cially the finding that mixture between highly differentiated populations is a recurrent process in the human past. Today, many people
assume that humans can be grouped biologically into "primeval"
groups, corresponding to our notion of "races," whose origins are
populations that separated tens of thousands of years ago. But this
long-held view about "race" has just in the last few years been proven
wrong—and the critique of concepts of race that the new data provide is very different from the classic one that has been developed
by anthropologists over the last hundred years. A great surprise that
emerges from the genome revolution is that in the relatively recent
past, human populations were just as different from each other as
they are today, but that the fault lines across populations were almost
unrecognizably different from today. DNA extracted from remains
of people who lived, say, ten thousand years ago shows that the structure of human populations at that time was qualitatively different.
Present-day populations are blends of past populations, which were
blends themselves. The African American and Latino populations of
the Americas are only the latest in a long line of major population
mixtures.

Who We Are and How We Got Here is divided into three parts.
Part I, "The Deep History of Our Species," describes how the
human genome not only provides all the information that a fertil-
ized human egg needs to develop, but also contains within it the
history of our species. Chapter 1, "How the Genome Explains Who
We Are," argues that the genome revolution has taught us about
who we are as humans not by revealing the distinctive features of our
biology compared to other animals but by uncovering the history
of migrations and population mixtures that formed us. Chapter 2,
"Encounters with Neanderthals," reveals how the breakthrough
technology of ancient DNA provided data from Neanderthals, our
big-brained cousins, and showed how they interbred with the ancestors of all modern humans living outside of Africa. The chapter also
explains how genetic data can be used to prove that ancient mixture
between populations occurred. Chapter 3, "Ancient DNA Opens
the Floodgates," highlights how ancient DNA can reveal features
of the past that no one had anticipated, starting with the discovery of
the Denisovans, a previously unknown archaic population that had

not been predicted by archaeologists and that mixed with the ancestors of present-day New Guineans. The sequencing of the Denisovan genome unleashed a cavalcade of discoveries of additional archaic populations and mixtures, and demonstrated unequivocally that population mixture is central to human nature.

Part II, "How We Got to Where We Are Today," is about how the genome revolution and ancient DNA have transformed our understanding of our own particular lineage of modern humans, and it takes readers on a tour around the world with population mixture as a unifying theme. Chapter 4, "Humanity's Ghosts," introduces the idea that we can reconstruct populations that no longer exist in unmixed form based on the bits of genetic material they have left behind in present-day people. Chapter 5, "The Making of Modern Europe," explains how Europeans today descend from three highly divergent populations, which came together over the last nine thousand years in a way that archaeologists never anticipated before ancient DNA became available. Chapter 6, "The Collision That Formed India," explains how the formation of South Asian populations parallels that of Europeans. In both cases, a mass migration of farmers from the Near East after nine thousand years ago mixed with previously established hunter-gatherers, and then a second mass migration from the Eurasian steppe after five thousand years ago brought a different kind of ancestry and probably Indo-European languages as well. Chapter 7, "In Search of Native American Ancestors," shows how the analysis of modern and ancient DNA has demonstrated that Native American populations prior to the arrival of Europeans derive ancestry from multiple major pulses of migration from Asia. Chapter 8, "The Genomic Origins of East Asians," describes how much of East Asian ancestry derives from major expansions of populations from the Chinese agricultural heartland. Chapter 9, "Rejoining Africa to the Human Story," highlights how ancient DNA studies are beginning to peel back the veil on the deep history of the African continent drawn by the great expansions of farmers in the last few thousand years that overran or mixed with previously resident populations.

Part III, "The Disruptive Genome," focuses on the implications of the genome revolution for society. It offers some suggestions for how to conceive of our personal place in the world, our connection to the more than seven billion people who live on earth with us, and the even larger numbers of people who inhabit our past and future. Chapter 10, "The Genomics of Inequality," shows how ancient DNA studies have revealed the deep history of inequality in social power among populations, between the sexes, and among individuals within a population, based on how that inequality determined success or failure of reproduction. Chapter 11, "The Genomics of Race and Identity," argues that the orthodoxy that has emerged over the last century—the idea that human populations are all too closely related to each other for there to be substantial average biological differences among them—is no longer sustainable, while also showing that racist pictures of the world that have long been offered as alternatives are even more in conflict with the lessons of the genetic data. The chapter suggests a new way of conceiving the differences among human populations—a way informed by the genome revolution. Chapter 12, "The Future of Ancient DNA," is a discussion of what comes next in the genome revolution. It argues that the genome revolution, with the help of ancient DNA, has realized Luca Cavalli-Sforza's dream, emerging as a tool for investigating past populations that is no less useful than the traditional tools of archaeology and historical linguistics. Ancient DNA and the genome revolution can now answer a previously unresolvable question about the deep past: the question of *what happened*—how ancient peoples related to each other and how migrations contributed to the changes evident in the archaeological record. Ancient DNA should be liberating to archaeologists because with answers to these questions in reach, archaeologists can get on with investigating what they have always been at least as interested in, which is *why* the changes occurred.

Before diving into the book, I will recount something that happened during a guest lecture I gave at the Massachusetts Institute of Technology in 2009. Mine was one of the last lectures of the term, meant to add spice to a course aimed at introducing students to computer-aided research into genomes with the goal of finding cures for disease. As I addressed Indian population history, an undergraduate sitting at the center of the front row stared me down. When I concluded, she asked me, with a grin, "How do you get funded to do this stuff?"

I mumbled something about how the human past shapes genetic variation, and about how, in order to identify risk factors for disease, it is important to understand that past. I gave the example of how among the thousands of distinctive human populations of India, there are high rates of disease because mutations that happened to be carried by the founders increased in frequency as the groups expanded. I make arguments along these lines in my applications to the U.S. National Institutes of Health, in which I propose to find disease risk factors that occur at different frequencies across populations. Grants of this type have funded much of my work since I started my laboratory in 2003.

True as these arguments are, I wish I had responded differently. We scientists are conditioned by the system of research funding to justify what we do in terms of practical application to health or technology. But shouldn't intrinsic curiosity be valued for itself? Shouldn't fundamental inquiry into who we are be the pinnacle of what we as a species hope to achieve? Isn't an attribute of an enlightened society that it values intellectual activity that may not have immediate economic or other practical impact? The study of the human past—as of art, music, literature, or cosmology—is vital because it makes us aware of aspects of our common condition that are profoundly important and that we heretofore never imagined.

Part I

The Deep History of Our Species

The Age of Modern Humans

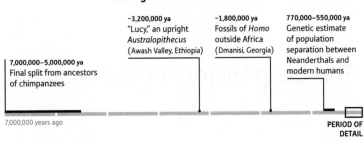

7,000,000–5,000,000 ya
Final split from ancestors
of chimpanzees

~3,200,000 ya
"Lucy," an upright
Australopithecus
(Awash Valley, Ethiopia)

~1,800,000 ya
Fossils of *Homo*
outside Africa
(Dmanisi, Georgia)

770,000–550,000 ya
Genetic estimate
of population
separation between
Neanderthals and
modern humans

7,000,000 years ago

PERIOD OF
DETAIL

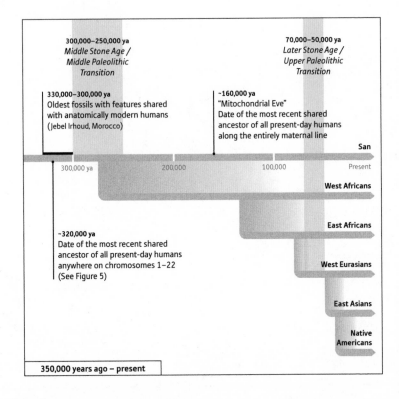

300,000–250,000 ya
*Middle Stone Age /
Middle Paleolithic
Transition*

70,000–50,000 ya
*Later Stone Age /
Upper Paleolithic
Transition*

330,000–300,000 ya
Oldest fossils with features shared
with anatomically modern humans
(Jebel Irhoud, Morocco)

~160,000 ya
"Mitochondrial Eve"
Date of the most recent shared
ancestor of all present-day humans
along the entirely maternal line

San

300,000 ya   200,000   100,000   Present

West Africans

East Africans

~320,000 ya
Date of the most recent shared
ancestor of all present-day humans
anywhere on chromosomes 1–22
(See Figure 5)

West Eurasians

East Asians

Native
Americans

350,000 years ago – present

1

How the Genome Explains Who We Are

The Master Chronicle of Human Variation

To understand why genetics is able to shed light on the human past, it is necessary to understand how the genome—defined as the full set of genetic code each of us inherits from our parents—records information. Francis Crick, Rosalind Franklin, James Watson, and Maurice Wilkins showed in 1953 that the genome is written out in twin chains of about three billion chemical building blocks (six billion in all) that can be thought of as the letters of an alphabet: A (adenine), C (cytosine), G (guanine), and T (thymine).[1] What we call a "gene" consists of tiny fragments of these chains, typically around one thousand letters long, which are used as templates to assemble the proteins that do most of the work in cells. In between the genes is noncoding DNA, sometimes referred to as "junk" DNA. The order of the letters can be read by machines that perform chemical reactions on fragments of DNA, releasing flashes of light as the reactions pass along the length of the DNA sequence. The reactions emit a different color for each of the letters A, C, G, and T, so that the sequence of letters can be scanned into a computer by a camera.

Although the great majority of scientists are focused on the biological information that is contained within the genes, there are also occasional differences between DNA sequences. These differences are due to random errors in copying of genomes (known as muta-

The genome can be understood as a sequence of letters.

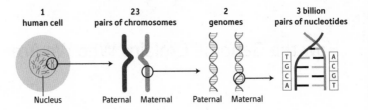

Differences in those sequences are caused by mutations.

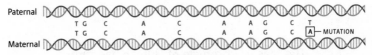

Figure 3. The genome contains about three billion nucleotides, which can be thought of as four letters in a biological alphabet: adenine (A), cytosine (C), guanine (G), and thymine (T). Around 99.9 percent of these letters are identical across two lined-up genomes, but in that last ~0.1 percent there are differences, reflecting mutations that accumulate over time. These mutations tell us how closely related two people are and record exquisitely precise information about the past.

tions) that occurred at some point in the past. It is these differences, occurring about one every thousand letters or so in both genes and in "junk," that geneticists study to learn about the past. Over the approximately three billion letters, there are typically around three million differences between unrelated genomes. The higher the density of differences separating two genomes on any segment, the longer it has been since the segments shared a common ancestor as the mutations accumulate at a more or less constant rate over time. So the density of differences provides a biological stopwatch, a record of how long it has been since key events occurred in the past.

The first startling application of genetics to the study of the past involved mitochondrial DNA. This is a tiny portion of the genome—only approximately 1/200,000th of it—which is passed down along the maternal line from mother to daughter to granddaughter. In 1987, Allan Wilson and his colleagues sequenced a few hundred letters of mitochondrial DNA from diverse people around the world. By comparing the mutations that were different among these sequences, he and his colleagues were able to reconstruct a family tree of maternal relationships. What they found is that the deepest branch of the tree—the branch that left the main trunk earliest—is found today only in people of sub-Saharan African ancestry, suggesting that the ancestors of modern humans lived in Africa. In contrast, all non-Africans today descend from a later branch of the tree.[2] This finding became an important part of the triumphant synthesis of archaeological and genetic and skeletal evidence that emerged in the 1980s and 1990s for the theory that modern humans descend from ancestors who lived in the last hundred thousand years or so in Africa. Based on the rate at which mutations are known to accumulate, Wilson and his colleagues estimated that the most recent African ancestor of all the branches, "Mitochondrial Eve," lived sometime after 200,000 years ago.[3] The best current estimate is around 160,000 years ago, although it is important to realize that like most genetic dates, this one is imprecise because of uncertainty about the true rate at which human mutations occur.[4]

The finding of such a recent common ancestor was exciting because it refuted the "multiregional hypothesis," according to which present-day humans living in many parts of Africa and Eurasia descend substantially from an early dispersal (at least 1.8 million years ago) of *Homo erectus*, a species that made crude stone tools and had a brain about two-thirds the size of ours. The multiregional hypothesis implied that descendants of *Homo erectus* evolved in parallel across Africa and Eurasia to give rise to the populations that live in the same places today. The multiregional hypothesis would therefore predict that there would be mitochondrial DNA sequences among present-day people that are separated by close to two million years, the age of the dispersal of *Homo erectus*. However, the genetic data was impossible to reconcile with this prediction. The fact that all people today share a common mitochondrial DNA ancestor about ten times more recently showed that humans today largely descend from a much later expansion from Africa.

Anthropological evidence pointed to a likely scenario for what occurred. The earliest human skeletons with "anatomically modern" features—defined as falling within the range of variation of all humans today with regard to having a globular brain case and other traits—date up to two hundred to three hundred thousand years ago and are all from Africa.[5] Outside of Africa and the Near East, though, there is no convincing evidence of anatomically modern humans older than a hundred thousand years and very limited evidence older than around fifty thousand years.[6] Archaeological evidence of stone tool types also points to a great change after around fifty thousand years ago, a period known to archaeologists of West Eurasia as the Upper Paleolithic, and to archaeologists of Africa as the Later Stone Age. After this time, the technology for manufacturing stone tools became very different, and there were changes in style every few thousand years, compared to the glacial earlier pace of change. Humans in this period also began to leave behind far more artifacts that revealed their aesthetic and spiritual lives: beads made of ostrich eggshells, polished stone bracelets, body paint made from red iron oxide, and the world's first representational art. The world's earliest known figurine is a roughly forty-thousand-year-old "lion-man" carved from a woolly mammoth tusk, found in Hohlenstein-Stadel in Germany.[7] The approximately thirty-thousand-year-old drawings of pre–ice age beasts, found on the walls of Chauvet Cave in France, even today are recognizable as transcendent art.

The dramatic acceleration of change in the archaeological record after around fifty thousand years ago was also reflected by evidence of population change. The Neanderthals, who had evolved in Europe by around four hundred thousand years ago and are considered "archaic" in the sense that their skeletal shape did not fall within present-day human variation, went extinct in their last hold-out of western Europe between about forty-one thousand and thirty-nine thousand years ago, within a few thousand years of the arrival of modern humans.[8] Population turnovers also occurred elsewhere in Eurasia, as well as in southern Africa, where there is evidence of abandonment of sites and the sudden appearance of Later Stone Age cultures.[9]

The natural explanation for all these changes was the spread of an anatomically modern human population whose ancestors included "Mitochondrial Eve," who practiced a sophisticated new culture, and who largely replaced the people who lived in each place before.

The Siren Call of the Genetic Switch

The finding that genetics could help to distinguish between competing hypotheses of human origins led in the 1980s and 1990s to exuberance about the power of the discipline to provide simple explanations. Some even wondered if genetics might be able to do more than provide a supporting line of evidence for the spread of modern humans from Africa and the Near East after around fifty thousand years ago. Perhaps genes could also be the cause of that spread, offering an explanation as simple and beautiful as the four-letter code written in DNA for the quickening pace of change in the archaeological record.

The anthropologist best known for embracing the idea that a genetic change might explain how we came to be behaviorally distinct from our predecessors was Richard Klein. He put forward the idea that the Later Stone Age revolution of Africa and the Upper Paleolithic revolution of western Eurasia, when recognizably modern human behavior burst into full flower after about fifty thousand years ago, were driven by the rise in frequency of a single mutation of a gene affecting the biology of the brain, which permitted the manufacture of innovative tools and the development of complex behavior.

According to Klein's theory, the rise in frequency of this mutation primed humans for some enabling trait, such as the ability to use conceptual language. Klein thought that prior to the occurrence of this mutation, humans were incapable of modern behaviors. Supporting his notion are examples among other species of a small number of genetic changes that have effected major adaptations, such as the five changes that are sufficient to turn the tiny ears of the Mexican wild grass teosinte into the huge cobs of corn that we buy in the supermarket today.[10]

Klein's hypothesis came under intense criticism almost as soon as he suggested it, most notably from the archaeologists Sally McBrearty and Alison Brooks, who showed that almost every trait that Klein considered to be a hallmark of distinctly modern human behavior was evident in the African and Near Eastern archaeological records tens of thousands of years before the Upper Paleolithic and Later Stone Age transitions.[11] But even if no single behavior was new, Klein had put his finger on something important. The intensification of evidence for modern human behavior after fifty thousand years ago is undeniable, and raises the question of whether biological change contributed to it.

One geneticist who came of age at this time of exuberance about the power of genetics to provide simple explanations for great mysteries was Svante Pääbo, who arrived in Allan Wilson's laboratory just after the "Mitochondrial Eve" discovery, and who would go on to invent much of the toolkit of the ancient DNA revolution and to sequence the Neanderthal genome. In 2002, Pääbo and his colleagues discovered two mutations in the gene *FOXP2* that seemed to be candidates for propelling the great changes that occurred after around fifty thousand years ago. The previous year, medical geneticists had identified *FOXP2* as a gene that, when mutated, produces an extraordinary syndrome whose sufferers have normal-range cognitive capabilities, but cannot use complex language, including most grammar.[12] Pääbo and his colleagues showed that the protein produced by the *FOXP2* gene has remained almost identical during the more than hundred million years of evolution separating chimpanzees and mice. However, two changes to the protein occurred on just the human lineage since it branched out of the common ancestral population of humans and chimpanzees, showing that the gene had evolved much more rapidly on the human lineage.[13] Later work by Pääbo and his colleagues found that engineered mice with the human versions of *FOXP2* are identical to regular mice in most respects, but squeak differently, consistent with the idea that these changes affect the formation of sounds.[14] These two mutations at *FOXP2* cannot have contributed to the changes after fifty thousand years ago, since Neanderthals shared them,[15] but Pääbo and his colleagues later identified a third mutation that is found in almost all present-day humans and that affects when and in what cells *FOXP2* gets turned into protein. This change is absent in Neanderthals, and thus is a candidate for contributing to the evolution of modern humans after their separation from Neanderthals hundreds of thousands of years ago.[16]

Regardless of how important *FOXP2* itself is in modern human biology, Pääbo cites the search for the genetic basis for modern human behavior as a justification for sequencing the genomes of archaic humans.[17] Between 2010 and 2013, when he led a series of studies that published whole-genome sequences from archaic humans like Neanderthals, Pääbo's papers highlighted an evolving list of about one hundred thousand places in the genome where nearly all present-day humans carry genetic changes that are absent in Neanderthals.[18] There are surely biologically important changes hiding in the list, but we are still only at the very beginning of the process of determining what they are, reflecting a more general problem that we are like kindergartners in our ability to read the genome. While we have learned to decode the individual words—as we know how the sequence of DNA letters gets turned into proteins—we still can't parse the sentences.

The sad truth is that it is possible to count on the fingers of two hands the examples like *FOXP2* of mutations that increased in frequency in human ancestors under the pressure of natural selection and whose functions we partly understand. In each of these cases, the insights only came from years of hand-to-hand combat with life's secrets by graduate students or postdoctoral scientists making engineered mice or fish, suggesting that it will take an evolutionary Manhattan Project to understand the function of each mutation that we have and that Neanderthals do not. This Manhattan Project of human evolutionary biology is one to which we as a species should commit ourselves. But even when it is carried out, I expect that the findings will be so complicated—with so many individual genetic changes contributing to what makes humans distinctive—that few people will find the answer comprehensible. While the scientific question is profoundly important, I expect that no intellectually elegant and emotionally satisfying molecular explanation for behavioral modernity will ever be found.

But even if studying just a few locations in the genome will not provide a satisfying explanation for how modern human behavior evolved, the great surprise of the genome revolution is the explanations it is starting to provide from another perspective—that of history. By comprehending the entire genome—by going beyond the tiny slice of the past sampled by our mitochondrial DNA and Y chromosome and embracing the story of our past told by the multiplicity of our ancestors that is written in the record of our whole genome—we have already begun to sketch out a new picture of how we got to be the way we are. This explanation based on migrations and population mixture is the subject of this book.

One Hundred Thousand Adams and Eves

When the journalist Roger Lewin in 1987 dubbed the common maternal ancestor of all people living today "Mitochondrial Eve," he evoked a creation story—that of a woman who was the mother of us all, and whose descendants dispersed throughout the earth.[19] The name captured the collective imagination, and is still used not only by the public but also by many scientists to refer to this common maternal ancestor. But the name has been more misleading than helpful. It has fostered the mistaken impression that all of our DNA comes from precisely two ancestors and that to learn about our history it would be sufficient to simply track the purely maternal line represented by mitochondrial DNA, and the purely paternal line represented by the Y chromosome. Inspired by this possibility, the National Geographic Society's "Genographic Project," beginning in 2005, collected mitochondrial DNA and Y-chromosome data from close to a million people of diverse ethnic groups. But the project was outdated even before it began. It has been largely recreational, and has produced few interesting scientific results. From the outset, it was clear that most of the information about the human past present in mitochondrial DNA and Y-chromosome data had already been mined, and that far richer stories were buried in the whole genome.

The truth is that the genome contains the stories of many diverse ancestors—tens of thousands of independent genealogical lineages, not just the two whose stories can be traced with the Y chromosome and mitochondrial DNA. To understand this, one needs to realize that beyond mitochondrial DNA, the genome is not one continuous sequence from a single ancestor but is instead a mosaic. Forty-six of the mosaic tiles, as it were, are chromosomes—long stretches of

DNA that are physically separated in the cell. A genome consists of twenty-three chromosomes, and because a person carries two genomes, one from each parent, the total number is forty-six.

But the chromosomes themselves are mosaics of even smaller tiles. For example, the first third of a chromosome a woman passes down to her egg might come from her father and the last two-thirds from her mother, the result of a splicing together of her father's and mother's copies of that chromosome in her ovaries. Females create an average of about forty-five new splices when producing eggs, while males create about twenty-six splices when producing sperm, for a total of about seventy-one new splices per generation.[20] So it is that as we trace each generation back further into the past, a person's genome is derived from an ever-increasing number of spliced-together ancestral fragments.

This means that our genomes hold within them a multitude of ancestors. Any person's genome is derived from 47 stretches of DNA corresponding to the chromosomes transmitted by mother and father plus mitochondrial DNA. One generation back, a person's genome is derived from about 118 (47 plus 71) stretches of DNA transmitted by his or her parents. Two generations back, the number of ancestral stretches of DNA grows to around 189 (47 plus 71 plus another 71) transmitted by four grandparents. Look even further back in time, and the additional increase in ancestral stretches of DNA every generation is rapidly overtaken by the doubling of ancestors. Ten generations back, for example, the number of ancestral stretches of DNA is around 757 but the number of ancestors is 1,024, guaranteeing that each person has several hundred ancestors from whom he or she has received no DNA whatsoever. Twenty generations in the past, the number of ancestors is almost a thousand times greater than the number of ancestral stretches of DNA in a person's genome, so it is a certainty that each person has not inherited any DNA from the great majority of his or her actual ancestors.

These calculations mean that a person's genealogy, as reconstructed from historical records, is not the same as his or her genetic inheritance. The Bible and the chronicles of royal families record who begat whom over dozens of generations. Yet even if the genealogies are accurate, Queen Elizabeth II of England almost certainly

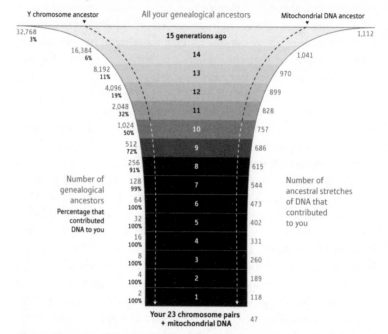

The Far Richer Story Told by the Whole Genome

Y chromosomes and mitochondrial DNA reflect information only
from the entirely male or entirely female lineages (dashed lines). The whole
genome carries information about tens of thousands of others.

Figure 4. The number of ancestors you have doubles every generation back in time. However, the number of stretches of DNA that contributed to you increases by only around seventy-one per generation. This means that if you go back eight or more generations, it is almost certain that you will have some ancestors whose DNA did not get passed down to you. Go back fifteen generations and the probability that any one ancestor contributed directly to your DNA becomes exceedingly small.

inherited no DNA from William of Normandy, who conquered England in 1066 and who is believed to be her ancestor twenty-four generations back in time.[21] This does not mean that Queen Elizabeth II did not inherit DNA from ancestors that far back, just that it is expected that only about 1,751 of her 16,777,216 twenty-fourth-degree genealogical ancestors contributed any DNA to her. This is such a small fraction that the only way William could plausibly be her genetic ancestor is if he was her genealogical ancestor in thousands of different lineage paths, which seems unlikely even considering the high level of inbreeding in the British royal family.

Going back deeper in time, a person's genome gets scattered into more and more ancestral stretches of DNA spread over ever-larger numbers of ancestors. Tracing back fifty thousand years in the past, our genome is scattered into more than one hundred thousand ancestral stretches of DNA, greater than the number of people who lived in any population at that time, so we inherit DNA from nearly everyone in our ancestral population who had a substantial number of offspring at times that remote in the past.

There is a limit, though, to the information that comparison of genome sequences provides about deep time. At each place in the genome, if we trace back our lineages far enough into the past, we reach a point where everyone descends from the same ancestor, beyond which it becomes impossible to obtain any information about deeper time from comparison of the DNA sequences of people living today. From this perspective, the common ancestor at each point in the genome is like a black hole in astrophysics, from which no information about deeper time can escape. For mitochondrial DNA this black hole occurs around 160,000 years ago, the date of "Mitochondrial Eve." For the great majority of the rest of the genome the black hole occurs between five million and one million years ago, and thus the rest of the genome can provide information about far deeper time than is accessible through analysis of mitochondrial DNA.[22] Beyond this, everything goes dark.

The power of tracing this multitude of lineages to reveal the past is extraordinary. In my mind's eye, when I think of a genome, I view it not as a thing of the present, but as deeply rooted in time, a tapestry of threads consisting of lines of descent and DNA sequences copied from parent to child winding back into the distant past. Tracing back, the threads wind themselves through ever more ancestors, providing information about population size and substructure in each generation. When an African American person is said to have 80 percent West African and 20 percent European ancestry, for example, a statement is being made that about five hundred years ago, prior to the population migrations and mixtures precipitated by European colonialism, 80 percent of the person's ancestral threads probably resided in West Africa and the remainder probably lived in Europe. But such statements are like still frames in a movie, capturing one point in the past. An equally valid perspective is that one hundred thousand years ago, the vast majority of lineages of African American ancestors, like those of everyone today, were in Africa.

The Story Told by the Multitudes in Our Genomes

In 2001, the human genome was sequenced for the first time—which means that the great majority of its chemical letters were read. About 70 percent of the sequence came from a single individual, an African American,[23] but some came from other people. By 2006, companies began selling robots that reduced the cost of reading DNA letters by more than ten thousandfold and soon by one hundred thousandfold, making it economical to map the genomes of many more people. It thus became possible to compare sequences not just from a few isolated locations, such as mitochondrial DNA, but from the whole genome. That made it possible to reconstruct each person's tens of thousands of ancestral lines of descent. This revolutionized the study of the past. Scientists could gather orders of magnitude more data, and test whether the history of our species suggested by the whole genome was the same as that told by mitochondrial DNA and the Y chromosome.

A 2011 paper by Heng Li and Richard Durbin showed that the idea that a single person's genome contains information about a multitude of ancestors was not just a theoretical possibility, but a reality. To decipher the deep history of a population from a single person's DNA, Li and Durbin leveraged the fact that any single person actually carries not one but two genomes: one from his or her father and one from his or her mother.[24] Thus it is possible to count the number of mutations separating the genome a person receives from his or her mother and the genome the person receives from his or her father to determine when they shared a common ancestor at each location. By examining the range of dates when these ancestors lived—plotting the ages of one hundred thousand Adams and

Eves—Li and Durbin established the size of the ancestral population at different times. In a small population, there is a substantial chance that two randomly chosen genome sequences derive from the same parent genome sequence, because the individuals who carry them share a parent. However, in a large population the chance is far lower. Thus, the times in the past when the population size was low can be identified based on the periods in the past when a disproportionate fraction of lineages have evidence of sharing common ancestors. Walt Whitman, in the poem "Song of Myself," wrote, "Do I contradict myself? / Very well, then I contradict myself, / (I am large, I contain multitudes)." Whitman could just as well have been talking about the Li and Durbin experiment and its demonstration that a whole population history is contained within a single person as revealed by the multitude of ancestors whose histories are recorded within that person's genome.

An unanticipated finding of the Li and Durbin study was its evidence that after the separation of non-African and African populations, there was an extended period in the shared history of non-Africans when populations were small, as reflected in evidence for many shared ancestors spread over tens of thousands of years.[25] A shared "bottleneck event" among non-Africans—when a small number of ancestors gave rise to a large number of descendants today—was not a new finding. But prior to Li and Durbin's work, there was no good information about the duration of this event, and it seemed plausible that it could have transpired over just a few generations—for example, a small band of people crossing the Sahara into North Africa, or from Africa into Asia. The Li and Durbin evidence of an extended period of small population size was also hard to square with the idea of an unstoppable expansion of modern humans both within and outside Africa around fifty thousand years ago. Our history may not be as simple as the story of a dominant group that was immediately successful wherever it went.

Figure 5

How We Can Tell How Long It Has Been
Since Our Genes Shared Common Ancestors

1 Each of us has two genomes: one from our mother, one from our father. Some segments are more alike than others. The more differences—or mutations—in a given segment, the longer it's been since the gene copies bequeathed to us by our parents shared a common ancestor.

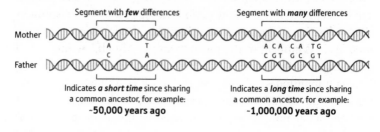

Segment with *few* differences   Segment with *many* differences

Mother

Father

Indicates *a short time* since sharing
a common ancestor, for example:
~50,000 years ago

Indicates a *long time* since sharing
a common ancestor, for example:
~1,000,000 years ago

2 For any pair of non-African genomes, more than 20% of individual genes share a common ancestor between 90,000 and 50,000 years ago. This reflects a population bottleneck when a small number of founders gave rise to many descendants outside Africa living today.

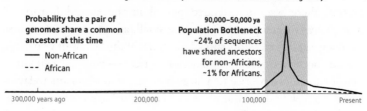

**Probability that a pair of
genomes share a common
ancestor at this time**

—— Non-African

- - - African

90,000–50,000 ya
Population Bottleneck
~24% of sequences
have shared ancestors
for non-Africans,
~1% for Africans.

300,000 years ago   200,000   100,000   Present

3 Across chromosomes 1–22, the most recent shared ancestor for all present-day people ranges mostly between 5,000,000 and 1,000,000 years ago, and nowhere is it estimated to be more recent than about 320,000 years ago.

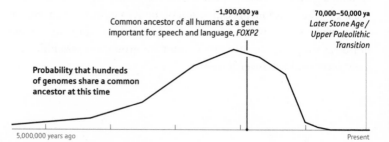

~1,900,000 ya
Common ancestor of all humans at a gene
important for speech and language, *FOXP2*

70,000–50,000 ya
*Later Stone Age /
Upper Paleolithic
Transition*

**Probability that hundreds
of genomes share a common
ancestor at this time**

5,000,000 years ago   Present

How the Whole-Genome Perspective
Put an End to Simple Explanations

The newfound ability to take a whole-genome view of human biology, made possible by leaps in technology in the last decades, has allowed reconstruction of population history in far more detail than had been previously possible. In doing so it revealed that the simple picture from mitochondrial DNA, and the just-so stories about one or a few changes propelling the Later Stone Age and Upper Paleolithic transitions when recognizably modern human behavior became widespread as reflected in archaeological sites across Africa and Eurasia, are no longer tenable.

In 2016, my colleagues and I used an adaptation of the Li and Durbin method[26] to compare populations from around the world to the earliest branching modern human lineage that has contributed a large proportion of the ancestry of a population living today: the one that contributed the lion's share of ancestry to the San hunter-gatherers of southern Africa. Our study,[27] like most others,[28] found that the separation had begun by around two hundred thousand years ago and was mostly complete by more than one hundred thousand years ago. The evidence for this is that the density of mutations separating San genomes from non-San genomes is uniformly high, implying few shared ancestors between San and non-San in the last hundred thousand years. "Pygmy" groups from Central African forests harbor ancestry that is arguably just as distinctive. The extremely ancient isolation of some pairs of human populations from each other conflicts with the idea that a single mutation essential to distinctively modern human behavior occurred shortly before the Upper Paleolithic and Later Stone Age. A key change essential to modern human behavior in this time frame would be expected to be at high frequency in some human populations today—those that descend from the population in which the mutation occurred—and absent or very rare in others. But this seems hard to reconcile with the fact that all people today are capable of mastering conceptual language and innovating their culture in a way that is a hallmark of modern humans.

A second problem with the notion of a genetic switch became apparent when we applied the Li and Durbin method to search for places where all the genomes we analyzed shared a common ancestor in the period before the Upper Paleolithic and Later Stone Age. At *FOXP2*—the gene that seemed the best candidate for a switch based on previous studies—we found that the common ancestor of everyone living today (that is, the person in whom modern humanity's shared copy of *FOXP2* last occurred), lived more than one million years ago.[29]

Expanding our analysis to the whole genome, we could not find any location—apart from mitochondrial DNA and the Y chromosome—where all people living today share a common ancestor less than about 320,000 years ago. This is a far longer time scale than the one required by Klein's hypothesis. If Klein was right, it would be expected that there would be places in the genome, beyond mitochondrial DNA and the Y chromosome, where almost everyone shares a common ancestor within the last hundred thousand years. But these do not in fact seem to exist.

Our results do not completely rule out the hypothesis of a single critical genetic change. There is a small fraction of the genome that contains complicated sequences that are difficult to study and that was not included in our survey. But the key change, if it exists, is running out of places to hide. The time scale of human genetic innovation and population differentiation is also far longer than mitochondrial DNA and other genetic data suggested prior to the genome revolution. If we are going to try to search the genome for clues to what makes modern humans distinctive, it is likely that we cannot look to explanations involving one or a few changes.

The whole-genome approaches that became possible after the technological revolution of the 2000s also soon made it clear that natural selection was not likely to take the simple form of changes in a small number of genes, as Klein had imagined. When the first whole-genome datasets were published, many geneticists (myself included) developed methods that scoured the genome for mutations that were affected by natural selection.[30] We were searching for the "low-hanging fruit"—instances in which natural selection had operated strongly on a few mutations. Examples of such low-hanging fruit include the mutations allowing people to digest cow's milk into adulthood, or mutations that cause darkening or lightening of skin to adapt to local climates, or mutations that bequeath resistance to the infectious disease malaria. As a community, we have been successful in identifying selection on mutations like these because they have risen rapidly from low to high frequency, resulting in a large number of people today sharing a recent ancestor or striking differences in mutation frequency between two otherwise similar populations. Events like these leave great scars on patterns of genome variation that can be detected without too much trouble.

Excitement about this bonanza was tempered by work led by Molly Przeworski, who studied the types of patterns that natural selection is likely to leave on the genome as a whole. A 2006 study by Przeworski and her colleagues showed that genome scans of present-day human genetic variation will miss most instances of natural selection because they simply will not have the statistical power needed to detect it, and that scans of this type will also have more power to detect some types of selection than others.[31] A study she led in 2011 then showed that only a small fraction of evolution in humans has likely involved intense natural selection for advantageous mutations that had not previously been present in the population.[32] Thus, intense and easily detectable episodes of natural selection such as those that have facilitated the digestion of cow's milk into adulthood are an exception.[33]

So what has been the dominant mode of natural selection in humans if not selection on newly arising single mutation changes that then rocket up to high frequency? An important clue comes from the study of height. In 2010, medical geneticists analyzed the genomes of around 180,000 people with measured heights, and found 180 independent genetic changes that are more common in shorter people. This means that these changes, or ones nearby on the genome, contribute directly to reduced height. In 2012, a second study showed that at the 180 changes, southern Europeans tend to have the versions that reduce height, and that this pattern is so pronounced that the only possible explanation is natural selection—likely for increased height in northern Europeans or decreased height in southern Europeans since the two lineages separated.[34] In

2015, an ancient DNA study led by Iain Mathieson in my laboratory revealed more about this process. We assembled DNA data from the bones and teeth of 230 ancient Europeans and analyzed the data to suggest that these patterns reflected natural selection for mutations that decreased height in farmers in southern Europe after eight thousand years ago, or increased height in ancestors of northern Europeans who lived in the eastern European steppe lands before five thousand years ago.[35] The advantages that accrued to shorter people in southern Europe, or to taller people in far eastern Europe, must have increased the number of their surviving children, which had the effect of systematically changing the frequencies of these mutations until a new average height was achieved.

Since the discoveries about height, other scientists have documented additional examples of natural selection on other complex human traits. A 2016 study analyzed the genomes of several thousand present-day Britons and found natural selection for increased height, blonder hair, bluer eyes, larger infant head size, larger female hip size, later growth spurt in males, and later age of puberty in females.[36]

These examples demonstrate that by leveraging the power of the whole genome to examine thousands of independent positions in the genome simultaneously, it is possible to get beyond the barrier that Molly Przeworski had identified—"Przeworski's Limit"—by taking advantage of information that we now have about a large number of genetic variations at many locations in the genome that have similar biological effects. We have such information from "genome-wide association studies," which since 2005 have collected data from more than one million people with a variety of measured traits, thereby identifying more than ten thousand individual mutations that occur at significantly elevated frequency in people with particular traits, including height.[37] The value of genome-wide association studies for understanding human health and disease has been contentious because the specific mutation changes that these studies have identified typically have such small effects that their results are hardly useful for predicting who gets a disease and who does not.[38] But what is often overlooked is that genome-wide association studies have provided a powerful resource for investigating human evolutionary change over time. By testing whether the mutations identified by genome-wide association studies as affecting particular biological traits have all tended to shift in frequency in the same direction, we can obtain evidence of natural selection for specific biological traits.

As genome-wide association studies proceed, they are beginning to investigate human variation in cognitive and behavioral traits,[39] and studies like these—such as the ones for height—will make it possible to explore whether the shift to behavioral modernity among our ancestors was driven by natural selection. This means that there is new hope for providing genetic insight into the mystery that puzzled Klein—the great change in human behavior suggested by the archaeological records of the Upper Paleolithic and Later Stone Age.

But even if genetic changes—through coordinated natural selection on combinations of many mutations simultaneously—did enable new cognitive capacities, this is a very different scenario from Klein's idea of a genetic switch. Genetic changes in this scenario are not a creative force abruptly enabling modern human behavior, but instead are responsive to nongenetic pressures imposed from the outside. In this scenario, it is not the case that the human population was unable to adapt because no one carried a mutation that allows a biological capability not previously present. Instead, the genetic formula that may have been necessary to drive the striking advances in human behavior and capacities that occurred during the Upper Paleolithic and Later Stone Ages is not particularly mysterious. The mutations necessary to facilitate modern human behavior were already in place, and many alternative combinations of these mutations could have increased in frequency together due to natural selection in response to changing needs imposed by the development of conceptual language or new environmental conditions. This in turn could have enabled further changes in lifestyle and innovation, in a self-reinforcing cycle. Thus, even if it is true that increases in the frequency of mutations were important in allowing modern humans to match their biology to new conditions during the Upper Paleolithic and Later Stone Age transition, what we now know about the nature of natural selection in humans and about the genetic encoding of many biological traits means it is unlikely that the first occurrence of these mutations triggered the great changes that followed. If we search for answers in a small number of mutations that arose shortly before the time of the Upper Paleolithic and Later Stone Age transitions, we are unlikely to find satisfying explanations of who we are.

How the Genome Can Explain Who We Are

It was molecular biologists who first focused the power of the genome onto the study of human evolution. Perhaps because of their background—and their track record of using reductionist approaches to solve great mysteries of life like the genetic code— molecular biologists were motivated by the hope that genetics would provide insights into the biological nature of how humans differ from other animals. Excitement about this prospect has also been shared by archaeologists and the public. But this research program, important as it is, is still at its very beginning because the answer is not going to be simple.

It is in the area of shedding light on human migrations—rather than in explaining human biology—that the genome revolution has already been a runaway success. In the last few years, the genome revolution—turbocharged by ancient DNA—has revealed that human populations are related to each other in ways that no one expected. The story that is emerging differs from the one we learned as children, or from popular culture. It is full of surprises: massive mixtures of differentiated populations; sweeping population replacements and expansions; and population divisions in prehistoric times that did not fall along the same lines as population differences that exist today. It is a story about how our interconnected human family was formed, in myriad ways never imagined.

The Age of Neandertals

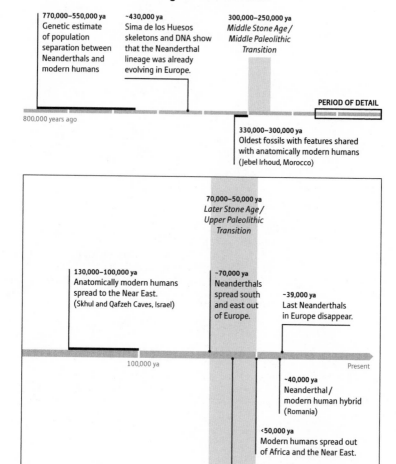

770,000–550,000 ya
Genetic estimate of population separation between Neanderthals and modern humans

~430,000 ya
Sima de los Huesos skeletons and DNA show that the Neanderthal lineage was already evolving in Europe.

300,000–250,000 ya
Middle Stone Age / Middle Paleolithic Transition

PERIOD OF DETAIL

800,000 years ago

330,000–300,000 ya
Oldest fossils with features shared with anatomically modern humans (Jebel Irhoud, Morocco)

70,000–50,000 ya
Later Stone Age / Upper Paleolithic Transition

130,000–100,000 ya
Anatomically modern humans spread to the Near East. (Skhul and Qafzeh Caves, Israel)

~70,000 ya
Neanderthals spread south and east out of Europe.

~39,000 ya
Last Neanderthals in Europe disappear.

100,000 ya

Present

~40,000 ya
Neanderthal / modern human hybrid (Romania)

<50,000 ya
Modern humans spread out of Africa and the Near East.

~60,000 ya
Neanderthal skeleton (Kebara Cave, Israel)

150,000 years ago – present

2

Encounters with Neanderthals

The Meeting of Neanderthals and Modern Humans

Today, the particular subgroup of humans to which we belong—modern humans—is alone on our planet. We outcompeted or exterminated other humans, mostly during the period after around fifty thousand years ago when modern humans expanded throughout Eurasia and when major movements of humans likely happened within Africa too. Today, our closest living relatives are the African apes: the chimpanzees, bonobos, and gorillas, all incapable of making sophisticated tools or using conceptual language. But until around forty thousand years ago, the world was inhabited by multiple groups of archaic humans who differed from us physically but walked upright and shared many of our capabilities. The question that the archaeological record cannot answer—but the DNA record can—is how those archaic people were related to us.

For no archaic group has the answer to this question seemed more urgent than for the Neanderthals. In Europe after four hundred thousand years ago, the landscape was dominated by these large-bodied people with brains slightly bigger on average than those of modern humans. The specimen that gave its name to Neanderthals was found in 1856 by miners in a limestone quarry in the Neander Valley (the German word for valley is *Thal* or *Tal*). For years, debate raged over whether these remains came from a deformed human, a human ancestor, or a human lineage that is extremely divergent from our own. Neanderthals became the first archaic humans to be recognized by science. In *The Descent of Man*, published in 1871, Charles Darwin argued that humans are like other animals in that they are also the products of evolution.[1] Although Darwin didn't himself appreciate their significance, Neanderthals were eventually acknowledged to be from a population more closely related to modern humans than to living apes, providing evidence for Darwin's theory that such populations must have existed in the past.

Over the next century and a half there were discoveries of many additional Neanderthal skeletons. These studies revealed that Neanderthals had evolved in Europe from even more archaic humans. In popular culture, they garnered a reputation as beastly—much more different from us than they in fact were. The primitive reputation of Neanderthals was fueled in large part by a slouched reconstruction of the Neanderthal skeleton from La Chapelle-aux-Saints, France, made in 1911. But from all the evidence we have, before about one hundred thousand years ago, Neanderthals were behaviorally just as sophisticated as our own ancestors—anatomically modern humans.

Both Neanderthals and anatomically modern humans made stone tools using a technique that has become known as Levallois, which requires as much cognitive skill and dexterity as the Upper Paleolithic and Later Stone Age toolmaking techniques that emerged among modern humans after around fifty thousand years ago. In this technique, flakes are struck off carefully prepared rock cores that have little resemblance to the resulting tools, so that craftspeople must hold in their minds an image of what the finished tool will look like and execute the complex steps by which the stone must be worked to achieve that goal.

Other signs of the cognitive sophistication of Neanderthals include the evidence that they cared for their sick and elderly. An excavation at Shanidar Cave in Iraq has revealed nine skeletons, all apparently deliberately buried, one of which was a half-blind elderly man with a withered arm, suggesting that the only way he could have survived is if friends and family had lovingly cared for him.[2] The Neanderthals also had an appreciation of symbolism, as revealed by jewelry made of eagle talons found at Krapina Cave in Croatia and dating to about

130,000 years ago,[3] and stone circles built deep inside Bruniquel Cave in France and dating to around 180,000 years ago.[4]

Yet despite similarities between Neanderthals and modern humans, profound differences are evident. An article written in the 1950s claimed that a Neanderthal on the New York City subway would attract no attention, "provided that he were bathed, shaved, and dressed in modern clothing."[5] But in truth, his or her strangely projecting brow and impressively muscular body would be give-aways. Neanderthals were much more different from any human population today than present-day populations are from each other.

The encounter of Neanderthals and modern humans has also cap-tured the imagination of novelists. In William Golding's 1955 *The Inheritors*, a band of Neanderthals is killed by modern humans, who adopt a surviving Neanderthal child.[6] In Jean Auel's 1980 *The Clan of the Cave Bear*, a modern human woman is brought up by Neander-thals, and the conceit of the book is a dramatization of what close interaction of these two sophisticated groups of humans, so alien to each other and yet so similar, might have been like.[7]

There is hard scientific evidence that modern humans and Nean-derthals met. The most direct is from western Europe, where Nean-derthals disappeared around thirty-nine thousand years ago.[8] The arrival of modern humans in western Europe was at least a few thou-sand years earlier, as is evident at Fumane in southern Italy where around forty-four thousand years ago, Neanderthal-type stone tools gave way to tools typical of modern humans. In southwest-ern Europe, tools typical of modern humans, made in a style called Châtelperronian, have been found amidst Neanderthal remains that date to between forty-four thousand and thirty-nine thousand years ago, suggesting that Neanderthals may have imitated modern human toolmaking, or that the two groups traded tools or materials. Not all archaeologists accept this interpretation, though, and there is ongo-ing debate about whether Châtelperronian artifacts were made by Neanderthals or by modern humans.[9]

Meetings between Neanderthals and modern humans took place not only in Europe but almost certainly in the Near East as well. After around seventy thousand years ago, a strong and successful Neanderthal population expanded from Europe into central Asia

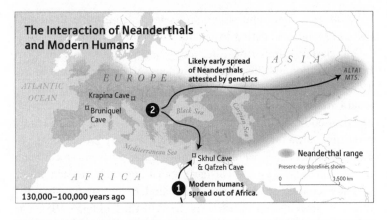

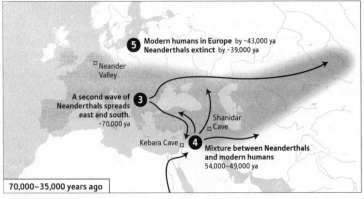

Figure 6. After around 400,000 years ago, Neanderthals were the dominant humans in western Eurasia, eventually extending as far east as the Altai Mountains. They survived an initial influx of modern humans at least by 120,000 years ago. Then, after 60,000 years ago, modern humans made a second push out of Africa into Eurasia. Before long, the Neanderthals went extinct.

as far as the Altai Mountains, and into the Near East. The Near East had already been inhabited by modern humans, as attested by remains at Skhul Cave on the Carmel Ridge in Israel and Qafzeh Cave in the Lower Galilee dating to between about 130,000 and 100,000 years ago.[10] Later, Neanderthals moved into the region, with one skeleton at Kebara Cave on the Carmel Ridge dating to between sixty and forty-eight thousand years ago.[11] Reversing the expectation we might have that modern humans displaced Neanderthals at every encounter, Neanderthals were advancing from their homeland (Europe) even as modern humans retreated. Sometime after sixty thousand years ago, though, modern humans began to predominate in the Near East. Now the Neanderthals were the losers in the encounter, and they went extinct not only in the Near East but eventually elsewhere in Eurasia as well. So it was that in the Near East there were at least two opportunities for encounters between Neanderthals and modern humans: when early modern humans first peopled the region before around one hundred thousand years ago and established a population that met the expanding Neanderthals, and when modern humans returned and displaced the Neanderthals there sometime around sixty or fifty thousand years ago.

Did the two populations interbreed? Are the Neanderthals among the direct ancestors of any present-day humans? There is some skeletal evidence for hybridization. Erik Trinkaus identified remains such as those from Oase Cave in Romania that he argued were intermediate between modern humans and Neanderthals.[12] However, shared skeletal features sometimes reflect adaptation to the same environmental pressures, not shared ancestry. This is why archaeological and skeletal records cannot determine the relatedness of Neanderthals to us. Studies of the genome can.

Neanderthal DNA

Early on, scientists studying ancient DNA focused almost exclusively on mitochondrial DNA, for two reasons. First, there are about one thousand copies of mitochondrial DNA in each cell, compared to two copies of most of the rest of the genome, increasing the chance of successful extraction. Second, mitochondrial DNA is information-dense: there are many more differences for a given number of DNA letters than in most other places in the genome, making it possible to obtain a more precise measurement of genetic separation time for every letter of DNA that is successfully analyzed. Mitochondrial data analysis confirmed that Neanderthals shared maternal-line ancestors with modern humans more recently than previously thought[13]—

the best current estimate is 470,000 to 360,000 years ago.[14] Mitochondrial DNA analysis also confirmed that the Neanderthals were highly distinctive. Their DNA type was outside the range of present-day variation in humans, sharing a common ancestor with us at a date several times more ancient than the time when "Mitochondrial Eve" lived.[15]

Neanderthal mitochondrial DNA provided no support for the theory that Neanderthals and modern humans interbred when they encountered each other, but at the same time the mitochondrial DNA evidence could not exclude up to around a 25 percent contribution of Neanderthals to the DNA of present-day non-Africans.[16] There is a reason why we have so little power to make statements about the Neanderthal contribution to modern humans based only on mitochondrial DNA. Even if modern humans outside Africa today do have substantial Neanderthal ancestry, there are only one or few women who lived at that time and were lucky enough to pass down their mitochondrial DNA to present-day people, and if most of those women were modern humans, the patterns we see today would not be surprising. So the mitochondrial data were not conclusive, but nevertheless the view that Neanderthals and modern humans did not mix remained the scientific orthodoxy until Svante Pääbo's team extracted DNA from the whole genome of a Neanderthal, making it possible to examine the history of all its ancestors, not just the exclusively maternal line.

The advance to sequencing the whole Neanderthal genome was made possible by a huge leap in the efficiency of the technology for studying ancient DNA in the decade after the sequencing of Neanderthal mitochondrial DNA.

The mainstay of ancient DNA research prior to 2010 was a technique called polymerase chain reaction (PCR). This involved selecting a stretch of DNA to be targeted, and then synthesizing approximately twenty-letter-long fragments of DNA that match the genome on each side of the targeted segment. These unique fragments pick out the targeted part of the genome, which is then duplicated many times over by enzymes. The effect is to take a tiny fraction of all the DNA in the sample and make it the dominant sequence. This method throws away the vast majority of DNA (the part that is not targeted). Nevertheless, it can extract at least some DNA that is of interest.

The new approach for extracting ancient DNA was radically different. It relied on sequencing all of the DNA in the sample, regardless of the part of the genome it comes from, and without preselecting the DNA based on targeting sequences. It took advantage of the brute power of new machines, which from 2006 to 2010 reduced the cost of sequencing by at least about ten thousandfold. The data could be analyzed by a computer to piece together most of a genome, or alternatively to pick out a gene of interest.

To make the new approach work, Pääbo's team needed to overcome several challenges. First, they needed to find a bone from which they could extract enough DNA. Anthropologists often work with fossils—bones completely mineralized into rocks. But it is impossible to get any DNA from a true fossil. Pääbo was therefore looking for bones that were not completely mineralized but contained organic material, including stretches of well-preserved DNA. Second, supposing the team could find a "golden sample" with well-preserved DNA, they still had to overcome the problem of contamination of the sample by microbial DNA, which comes from the bacteria and fungi that embed themselves in bone after an individual's death. These contribute the overwhelming majority of DNA in most ancient samples. Finally, the team had to consider the likelihood of contamination by the researchers—archaeologists or molecular biologists—who handled the samples and chemicals and may have left traces of their own DNA on them.

Contamination is a huge danger for studies of ancient human DNA. Contaminated sequences can mislead analysts because the modern humans handling the bone are related, even if very distantly, to the individual being sequenced. A typical Neanderthal ancient DNA fragment from a well-preserved sample is only about forty letters long, while the rate of differences between modern humans and Neanderthals is about one per six hundred letters, so it is sometimes impossible to tell whether a particular stretch of DNA comes from the bone or from someone who handled it. Contamination has bedeviled ancient DNA researchers time and again. For example, in 2006 Pääbo's group sequenced about a million letters of DNA from

Neanderthals as a trial run prior to whole-genome sequencing.[17] A high fraction of the sequences were modern human contaminants, compromising interpretation of the data.[18]

Modern measures to minimize the possibility of contamination in ancient DNA analysis, which had already begun to be implemented in the 2006 study and which became even more elaborate afterward, involve an obsessive set of precautions. For the 2010 study in which Pääbo and his team successfully sequenced an uncontaminated Neanderthal genome, they took each of the bones they screened into a "clean room," which they adapted from the blueprints of the clean spaces used in microchip fabrication facilities in the computer industry. There was an overhead ultraviolet (UV) light of the same type used in surgical operating suites that was turned on whenever researchers were not present, in order to convert contaminating DNA into a form that cannot be sequenced (the light also destroys ancient DNA on the outside of samples, but researchers drill beneath the surface and so are able to access DNA that is not destroyed). The air was ultra-filtered to remove tiny dust particles—anything more than one thousand times smaller than the width of a human hair—that might contain DNA. The suite was pressurized so that air flowed from inside to outside, to protect the samples from any contaminating DNA wafting in from outside the lab.

There were three separate rooms in the suite. In the first, the researchers donned full-body clean suits, gloves, and face masks. In the second, they placed the bones chosen for sampling into a chamber where they exposed to high-energy UV radiation, again with the goal of converting the contaminating DNA that might be lying on the surface into a form that cannot be sequenced. The researchers then cored the bones using a sterilized dental drill, collected tens or hundreds of milligrams of powder onto UV-irradiated aluminum foil, and deposited this powder into a UV-irradiated tube. In the third chamber, they immersed the powder into chemical solutions that removed bone minerals and protein, and ran the solution over pure sand (silicon dioxide), which under the right conditions binds the DNA while removing the compounds that poison the chemical reactions used for sequencing.

The researchers then transformed the resulting DNA fragments into a form that could be sequenced. First, they chemically removed the ragged ends of the DNA fragments that had been degraded after tens of thousands of years buried under the ground. In an extra measure to remove contamination beyond what had been done in the 2006 study, Pääbo and his team attached an artificially synthesized sequence of letters, a chemical "barcode," to the ends of the DNA fragments. Any contaminating sequences that entered the experiment after the attachment of the barcode could thus be distinguished from the DNA of the ancient sample. The final step was to attach molecular adapters at either end that allowed the DNA fragment to be sequenced in one of the new machines that had made sequencing tens of thousands of times cheaper than the previous technology.

The best-preserved Neanderthal samples turned out to be three approximately forty-thousand-year-old arm and leg bones from Vindija Cave in the highlands of Croatia. After sequencing from these bones, Pääbo's team found that the great majority of DNA fragments they obtained were from bacteria and fungi that had colonized the bones. But by comparing the millions of fragments to the present-day human and chimpanzee genome sequences, they found gold amidst the dross. These reference genomes were like the picture on a jigsaw puzzle box, providing the key to aligning the tiny fragments of DNA they had sequenced. The bones contained as much as 4 percent archaic human DNA.

Once Pääbo realized in 2007 that he would be able to sequence almost the entire Neanderthal genome, he assembled an international team of experts with the goal of ensuring that the analysis would do justice to the data. This is how I got involved, together with my chief scientific partner, the applied mathematician Nick Patterson. Pääbo reached out to us because over the previous five years we had established ourselves as innovators in the area of studying population mixture. Over the course of many trips to Germany, I played an important role in the analyses that proved interbreeding between Neanderthals and some modern humans.

Affinities Between Neanderthals and Non-Africans

The Neanderthal genome sequences we were working with were unfortunately full of errors. We could see as much because the data suggested that several times more mutations had occurred on the Neanderthal lineage than on the modern human lineage after the two sequences separated from their common ancestors. Most of these apparent mutations could not be real, since mutations occur at an approximately constant rate over time, and as the Neanderthal bones were ancient, they were actually closer in time to the common ancestor than are present-day human genomes, and so should have accumulated fewer mutations. Based on the degree of excess mutations on the Neanderthal lineage, we estimated that the Neanderthal sequences we were working with had a mistake approximately every two hundred DNA letters. While this might sound small, it is actually much higher than the rate of true differences between Neanderthals and present-day humans, so most of the differences we found between the Neanderthal sequence and present-day human sequences were errors created by the measurement process and not genuine differences between the Neanderthal and present-day human genomes. To deal with the problem, we restricted our study to positions in the genome that are known to be variable among present-day humans. At these positions, an error rate of about 0.5 percent was too low to confuse the interpretation. Based on these positions, we designed a mathematical test for measuring whether Neanderthals were more closely related to some present-day humans than to others.

The test we developed is now called the "Four Population Test," and it has become a workhorse for comparing populations. The test takes as its input the DNA letters seen at the same position in four genomes: for example, two modern human genomes, the Neanderthal, and a chimpanzee. It examines whether, at positions where there is a mutation distinguishing the two modern human genomes that is also observed in the Neanderthal genome—which must reflect a mutation that occurred prior to the final separation of

The Four Population Test

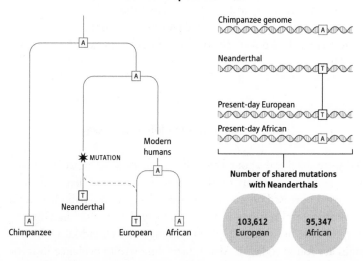

Figure 7. We can evaluate whether two populations are consistent with descending from a common ancestral population through the "Four Population Test." For example, consider a mutation that occurred in the ancestors of the Neanderthal (letter T, above) that is not seen in chimpanzee DNA. There are about 9 percent more of these mutations shared with Europeans than with African genomes, reflecting a history of Neanderthal interbreeding into the ancestors of Europeans.

Neanderthals and modern humans—the Neanderthal matches the second human population at a different rate from the first. If the two modern humans descend from a common ancestral population that separated earlier from the ancestors of Neanderthals, there is no reason why the mutation is more likely to have been passed down one modern human line than another, and thus the rate of matching of each of the two modern human genomes to Neanderthal is expected to be equal. In contrast, if Neanderthals and some modern humans interbred, the modern human population descended from the interbreeding will share more mutations with Neanderthals.

When we tested diverse present-day human populations, we found Neanderthals to be about equally close to Europeans, East Asians, and New Guineans, but closer to all non-Africans than to all sub-Saharan Africans, including populations as different as West Africans and San hunter-gatherers from southern Africa. The difference was slight, but the probability of these findings happening by chance was less than one in a quadrillion. We reached this conclusion however we analyzed the data. This was the pattern that would be expected if Neanderthals had interbred with the ancestors of non-Africans but not Africans.

Trying to Make the Evidence Go Away

We were skeptical about this conclusion because it went against the scientific consensus of the time—a consensus that had been strongly impressed on many members of our team. Pääbo had done his postdoctoral training in the laboratory that in 1987 had discovered that the most deeply splitting human mitochondrial DNA lineages are found today in Africa, providing strong evidence in favor of an African origin for all modern humans. Pääbo's own 1997 work strengthened the evidence for a purely African origin by showing that Neanderthal mitochondrial DNA fell far outside all modern human variation.[19]

I too came into the Neanderthal genome project with a strong bias against the possibility of Neanderthal interbreeding with modern humans. My Ph.D. supervisor, David Goldstein, was a student of Luca Cavalli-Sforza, who had made a fully out-of-Africa model a centerpiece of his models of human evolution, and I was steeped in this paradigm. The genetic data I knew about supported the out-of-Africa picture so consistently that from my perspective the strictest possible version of the out-of-Africa hypothesis, in which there was no interbreeding between the ancestors of present-day humans and Neanderthals, seemed like a good bet.

Coming from this background, we were deeply suspicious of the evidence we were finding for interbreeding with Neanderthals, and so we applied a particularly stringent series of tests in order to find some problem with our evidence. We tested whether the result was dependent on the genome sequencing technology that we used, but we obtained the same result from two very different technologies.

We considered the possibility that the finding might be an artifact of a high rate of error in ancient DNA, which is known to affect particular DNA letters much more than others. However, we obtained the same result regardless of the type of mutation we analyzed. We wondered if our finding resulted from contamination of the Neanderthal sample by present-day humans. This could perhaps have tainted the data despite the measures that Pääbo's team had taken to guard against it in the lab, and despite the tests we had performed on the data to measure the degree of modern human contamination, which had suggested that any contamination that was present was too small to produce the patterns we observed. However, even if there had been contamination from present-day humans, the patterns we observed looked nothing like what would be expected from it. If there had been contamination, it would most likely have come from a European, since almost all the Neanderthal bones we analyzed were excavated and handled by Europeans. Yet the Neanderthal sequence we had was no closer to Europeans than to East Asians or to New Guineans—three very different populations.

We remained skeptical, wondering if something we had not thought of could explain the patterns. Then, in June 2009, I attended a conference at the University of Michigan where I met Rasmus Nielsen, who had been scanning through the genomes of diverse humans from around the world. In most parts of the genome, Africans are more genetically diverse than non-Africans and carry the most deeply diverging lineages, as is the case with mitochondrial DNA. But Nielsen was identifying rare places in the genome where the genetic diversity among non-Africans was greater than in Africans because of lineages that split off the tree of present-day human sequences early and were present only in non-Africans. These sequences just might be derived from archaic humans who had interbred with non-Africans. Nielsen joined our collaboration and compared the regions he and his colleagues identified to the data. When he compared twelve of his special regions to the Neanderthal genome sequence, he found that in ten of them there was a close match to the Neanderthal. This was far too high a fraction to happen by chance. Most of Nielsen's highly divergent bits of DNA had to be Neanderthal in origin.

The chopping up of chromosomes every generation . . .

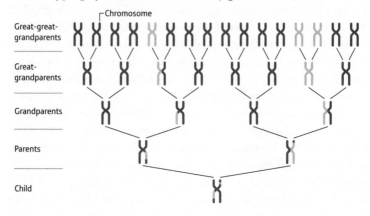

Next, we obtained a date for when the Neanderthal-related genetic material entered the ancestors of non-Africans. To do this, we took advantage of recombination—the process that occurs during the production of a person's sperm or eggs that swaps large segments of parental DNA to produce novel spliced chromosomes that are passed to the offspring. For example, consider a woman who is a first-generation mixture of a Neanderthal mother and a modern human father. In her cells, each pair of her chromosomes consists of one unbroken Neanderthal chromosome and one unbroken modern human chromosome. However, her eggs contain twenty-three mixed chromosomes. One chromosome in an egg of hers might have its first half of Neanderthal origin and its other half of modern human origin. Suppose she mates with a modern human, and mixture continues down the generations with more modern humans. Over the generations, the segments of Neanderthal DNA get chopped into smaller and smaller bits, with recombination operating like the whirring blade of a food processor, splicing the parental DNA at random positions along the chromosome in each generation. By measuring the typical sizes of the stretches of Neanderthal-related DNA in present humans, evident from the size of sequences that match the Neanderthal genome more than they do sub-Saharan African

... provides a clock for dating mixture events.

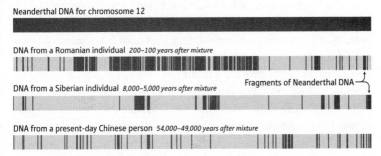

Neanderthal DNA for chromosome 12

DNA from a Romanian individual *200–100 years after mixture*

DNA from a Siberian individual *8,000–5,000 years after mixture*   Fragments of Neanderthal DNA

DNA from a present-day Chinese person *54,000–49,000 years after mixture*

Figure 8. When a person produces a sperm or an egg, he or she passes down to the next generation only one chromosome from each of the twenty-three pairs he or she carries. The transmitted chromosomes are spliced-together versions of the ones inherited from the mother and father (facing page). This means that the sizes of the bits of Neanderthal DNA in modern human genomes became smaller as the time since mixture increased (above, real data from chromosome 12).

genomes, we can learn how many generations have passed since the Neanderthal DNA entered a modern person's ancestors.

With this approach, we found that at least some Neanderthal-related genetic material came into the ancestors of present-day non-Africans eighty-six thousand to thirty-seven thousand years ago.[20] We have since refined this date by analyzing ancient DNA from a modern human from Siberia who, radiocarbon dating studies show, lived around forty-five thousand years ago. The stretches of Neanderthal-derived DNA in this individual are on average seven times larger than the stretches of Neanderthal-derived DNA in modern humans today, confirming that he lived much closer to the time of Neanderthal mixture. His proximity in time to the mixing event makes it possible to obtain a more accurate date of fifty-four thousand to forty-nine thousand years ago.[21]

But in 2012 we hadn't yet proven that the interbreeding we had detected was with Neanderthals themselves. The most serious questioning came from Graham Coop, who was convinced that we had detected interbreeding with archaic humans, but pointed out that it was possible that the interbreeding hadn't actually been with Neanderthals.[22] Instead, the patterns could be the result of interbreeding with an as yet unknown archaic human in turn distantly related to Neanderthals.

A year later we were able to rule out Coop's scenario after Pääbo's laboratory sequenced a high-quality Neanderthal genome from a toe bone found in southern Siberia dating to at least fifty thousand years ago (if a sample is older than about fifty thousand years, radiocarbon dating can only provide a minimum date, so it actually could be substantially older).[23] For this genome, we were able to gather about forty times more data than from the Croatian Neanderthal. With so much data, we could cross-check the sequence and edit away the errors. The resulting sequence was freer of errors than most genomes that are generated from living humans. The high-quality sequence allowed us to determine how closely related modern humans and Neanderthals are to each other based on the number of mutations that have occurred on the lineages since they separated. We found few or no segments where the Siberian Neanderthal shared common ancestors with present-day sub-Saharan Africans within the last half million years. However, there were shared segments with non-Africans roughly within the past one hundred thousand years. These dates fell within the time frame when Neanderthals were fully established in West Eurasia. This meant that the interbreeding was with true Neanderthals, not some distantly related groups.

Mixing in the Near East

So how much Neanderthal ancestry do people outside of Africa carry today? We found that non-African genomes today are around 1.5 to 2.1 percent Neanderthal in origin,[24] with the higher numbers in East Asians and the lower numbers in Europeans, despite the fact that Europe was the homeland of the Neanderthals.[25] We now know that at least part of the explanation is dilution. Ancient DNA from Europeans who lived before nine thousand years ago shows that pre-farming Europeans had just as much Neanderthal ancestry as East Asians do today.[26] The reduction in Neanderthal ancestry in present-day Europeans is due to the fact that they harbor some of their ancestry from a group of people who separated from all other non-Africans prior to the mixture with Neanderthals (the story of this early-splitting group revealed by ancient DNA is told in part II of this book). The spread of farmers with this inheritance diluted the Neanderthal ancestry in Europe, but not in East Asia.[27]

Based on archaeological evidence alone, it would seem a natural guess that Neanderthals interbred with modern humans in Europe, the place where Neanderthals originated. But is that the place where the main interbreeding that left its mark in people today occurred? The genetic data cannot tell us for sure. Genetic data can show how people are related, but humans are capable of migrating thousands of kilometers in a lifetime even on foot, so genetic patterns need not reflect events that occurred near the locations where the people who carry the DNA live. If the ancient DNA studies of the last few years have shown anything clearly, it is that the geographic distribution of people living today is often misleading about the dwelling places of their ancestors.

However, we can make plausible conjectures about geographic origin. Evidence of interbreeding is detected today not just in Europeans but also in East Asians and New Guineans. Europe is a cul-de-sac of sorts within Eurasia, and would not have been a likely detour for modern humans expanding eastward. So where could Neanderthals and modern humans have met and mixed to give rise to a population that expanded not only to Europe but also to East Asia and New Guinea? Archaeologists have shown how in the Near East, Neanderthals and modern humans traded places as the dominant human population at least twice between 130,000 and 50,000 years ago, and it is reasonable to guess that they might have met during this period. So interbreeding in the Near East provides a plausible explanation for the Neanderthal ancestry that is shared by Europeans and East Asians.

Did interbreeding happen in Europe at all? In 2014, Pääbo's group sequenced DNA from a skeleton from Oase Cave in Romania, the same skeleton that Erik Trinkaus had interpreted as a hybrid of Neanderthals with modern humans, based on features of its skull that were similar to both.[28] Our analysis of the data showed that the Oase individual, who radiocarbon dating studies had shown lived about forty thousand years ago, had around 6 to 9 percent Neanderthal ancestry, far more than the approximately 2 percent that we measure in present-day non-Africans.[29] Some stretches of Neanderthal DNA extend a third of the length of his chromosomes—a span so large and unbroken by recombination that we can be sure that the Oase individual had an actual Neanderthal no more than six generations back in his family tree. Contamination cannot explain these findings, as it would dilute the Neanderthal ancestry in the Oase individual, not increase it. It would also generate random matching to Neanderthals throughout the genome, not large stretches of Neanderthal DNA that could be readily identified by eye when we simply plotted along the genome the positions of mutations that match the Neanderthal genome sequence more closely than they match modern humans. This evidence of Neanderthal interbreeding didn't need statistics. The proof was in the picture.

The discoveries about the interbreeding in the recent family tree of the Oase individual suggested that modern humans and Neanderthals also hybridized in Europe, the homeland of the Neanderthals. But the population of which Oase was a part—and which carried this clear imprint of interbreeding with European Neanderthals—may not have left any descendants among people living today. When we analyzed the genome of Oase, we found no evidence that he was more closely related to Europeans than to East Asians. This means that he had to have been part of a population that was an evolutionary dead end—a pioneer modern human population that arrived early in Europe, flourished there briefly and interbred with local Neanderthals, and then went extinct. Thus, while the Oase individual provides powerful evidence that interbreeding between Neanderthals and modern humans occurred in Europe, he does not provide any evidence that Neanderthal ancestry in non-Africans today is derived from European Neanderthals. It remains the case that the most likely source of Neanderthal ancestry in non-Africans is Near Eastern Neanderthals.

The finding that Oase was from a dead-end population accords with the archaeological record of the first modern humans of Europe. The stone tools these humans made came in a variety of styles, but like the population of Oase himself, most were dead ends in the sense that they disappeared from the archaeological record after a few thousand years. However, one style known as the Protoaurignacian—thought to derive from the earlier Ahmarian of the Near East—persisted after thirty-nine thousand years ago and likely developed into the Aurignacian, the first widespread modern human culture in Europe.[30] These patterns could be explained if the makers of Aurignacian tools derived from a different migration into Europe compared to other early modern humans like Oase. This scenario could explain how it could be that Oase's population interbred heavily with local European Neanderthals, and yet the Neanderthal ancestry in Europeans today is not from Europe.

Two Groups at the Edge of Compatibility

The low fertility of hybrids may also have reduced Neanderthal ancestry in the DNA of people living today. This possibility was first advanced by Laurent Excoffier, who knew from studies of animals and plants that when one population moves into a region occupied by another population with which it can interbreed, even a small rate of interbreeding is enough to produce high proportions of mixture in the descendants—far more than the approximately 2 percent Neanderthal ancestry seen in non-Africans today. Excoffier argued that the only way that the modern human genome could have ended up with so little Neanderthal ancestry was if expanding modern humans had offspring with other modern humans at least fifty times more often than they did with the Neanderthals living in their midst.[31] He thought that the most likely explanation for this was that Neanderthals and modern human offspring were much less fertile than the offspring of matings between pairs of modern humans.

I wasn't convinced by this argument. Rather than low hybrid fertility, I favored the explanation that there simply wasn't much interbreeding for social reasons. Even today, many groups of modern humans keep largely to themselves because of cultural, religious, or caste barriers. Why should it have been any different for modern humans and Neanderthals when they encountered one another?

But Excoffier got something important right. This became evident when we and others analyzed the bits of Neanderthal DNA that entered into the modern human population and mapped their positions in the genome. To do this, Sriram Sankararaman in my laboratory searched for mutations that were present in the sequenced Neanderthals but were rare or absent in sub-Saharan Africans. By studying stretches of such mutations, we were able to find a substantial fraction of all the Neanderthal ancestry fragments in each non-African. Looking at where in the genome these Neanderthal ancestry fragments occurred, it became clear that the impact of Neanderthal interbreeding varied dramatically across the genome of non-African people today. The average proportion of Neanderthal ancestry in non-African populations is around 2 percent, but it is not spread evenly. In more than half the genome, no Neanderthal ancestry has been detected in anyone. But in some unusual places in the genome, more than 50 percent of DNA sequences are from Neanderthals.[32]

A critical clue that helped us to understand how this pattern had formed came from studying the places in non-African genomes where Neanderthal ancestry is rare. In any one stretch of DNA, an absence of Neanderthal ancestry in the population can happen by chance, as we think is the case for mitochondrial DNA. However, it is improbable that a substantial subset of the genome with particular biological functions will be systematically depleted of Neanderthal ancestry unless natural selection systematically worked to remove it.

But evidence of systematic removal of Neanderthal ancestry is exactly what we found—and, remarkably, we found a particularly intense depletion of Neanderthal ancestry by natural selection in two parts of the genome known to be relevant to the fertility of hybrids.

The first place of reduced Neanderthal ancestry was on chromosome X, one of the two sex chromosomes. This reminded me of a pattern that Nick Patterson and I had run into in our work on the separation of human and chimpanzee ancestors in a study we had carried out together and published years before.[33] There are only three copies of chromosome X in any population for every four other chromosomes (because females carry two copies and males only one, in contrast to two copies in each sex for most of the rest of the chromosomes). This means that in any one generation, the probability that any two X chromosomes share a common parent is four-thirds the probability that any two of one of the other chromosomes share a common parent. It follows that the expected time since any pair of X chromosome sequences descend from a common ancestral sequence is about three-quarters of that in the rest of the genome. In fact, though, the real data suggest a number that is around half or even less.[34] In our study of the common ancestral population of humans and chimpanzees, we had not been able to identify any history that could explain this pattern, such as a lower rate of females moving among groups than males, or a more variable number of children in females than in males, or population expansion or contraction. However, the patterns could be explained by a history in which the ancestors of humans and chimpanzees initially separated, then came together to form either human or chimpanzee ancestors before the final separation of the two lineages.

How is it that hybridization can lead to so much less genetic variation on chromosome X than on the rest of the genome? From studies of a variety of species across the animal kingdom, it is known that when two populations are separated for long enough, hybrid offspring have reduced fertility. In mammals like us, reduced fertility is much more common in males, and the genetic factors contributing to this reduced fertility are concentrated on chromosome X.[35] So when two populations are so separated that their offspring have reduced fertility, but nevertheless mix together to produce hybrids, it is expected that there will be intense natural selection to remove the factors contributing to reduced fertility. This process will be especially evident on chromosome X because of the concentration of genes contributing to infertility on it. As a result, there tends to be natural selection on chromosome X for stretches of DNA from the population that contributed most of the hybrid population's ancestry. This causes the hybrid population to derive its chromosome X almost entirely from the majority population, leading to an anomalously low genetic divergence on chromosome X between the hybrid population and one of the hybridizing populations, consistent with the pattern seen in humans and chimpanzees.

This theoretical prediction might sound fanciful, but in fact it is borne out in hybrids of the western European and eastern European house mouse species in a band of territory that runs in a north-to-south direction through central Europe, roughly along the line of the former Cold War Iron Curtain. While the density of mutations separating the hybrid mice from western European mice is high in most of the genome because the hybrid mice carry DNA not just from western European mice but also from highly divergent eastern European mice, the density on the X chromosome is far less because the hybrid mice harbor very little DNA from the eastern European population whose X chromosomes are known to cause infertility in male hybrids.[36]

Since the publication of our paper in 2006 suggesting that either humans or chimpanzees may derive from an ancient major hybridization, the evidence for ancient major hybridization in the ancestry of humans and chimpanzees has, if anything, become even stronger. In 2012 Mikkel Schierup, Thomas Mailund, and colleagues developed a new method to estimate the suddenness of separation of the ancestors of two present-day species from genetic data, based on principles similar to the Li and Durbin approach described in chapter one.[37] When they applied the method to study the separation time of common chimpanzees and their distant cousins, bonobos, they found evidence that the separation was very sudden, consistent with the hypothesis that the species were separated by a huge river (the Congo) that formed rather suddenly one to two million years ago. In contrast, when they applied the method to study humans and chimpanzees, they found evidence for an extended period of genetic interchange after population differentiation began, as expected for hybridization.[38]

An even more important piece of evidence came from a paper Schierup and Mailund published in 2015, when together with other colleagues, they showed that the regions that are denuded of Neanderthal mixture on chromosome X in non-Africans are to a large extent the same regions that are driving the low genetic divergence between humans and chimpanzees.[39] This is what would be expected if mutations that contribute to reduced fertility when they occur in a hybrid individual tend to be concentrated not just on chromosome

Figure 9

**Neanderthal ancestry has been removed
over time by natural selection.**

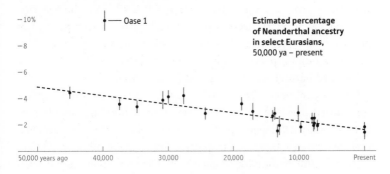

X, but in particular regions along chromosome X, causing the minority ancestry to be removed from the population by natural selection against the male hybrids who carry it. The evidence of selection to remove Neanderthal DNA from chromosome X was a tell-tale sign that male hybrids had reduced fertility.

We also found a second line of evidence for infertility in hybrids of Neanderthals and modern humans—a line of evidence that had nothing to do with the X chromosome. When reduced fertility is observed in hybrid males, the genes responsible tend to be highly active in the male reproductive tissue, causing malfunctions of sperm. So a prediction of the hypothesis of male hybrid infertility suggested to me by evolutionary biologist Daven Presgraves after I showed him the X chromosome evidence is that genes unusually active in the germ cells of a man's testicles will have less Neanderthal ancestry on average than genes that are most active in other body tissues. When we looked in real data, Presgraves's prediction was exactly borne out.[40]

The problems faced by modern humans with Neanderthal ancestry went beyond reduced fertility, as it turns out that Neanderthal ancestry is not just reduced on the X chromosome and around genes important in male reproduction, but is also reduced around the great majority of genes (there is far more Neanderthal ancestry in "junk" parts of the genome with few biological functions). The clearest evidence for this came from a study in 2016, in which we published a genome-wide ancient DNA dataset from more than fifty Eurasians spread over the last forty-five thousand years.[41] We showed that Neanderthal ancestry decreased continually from 3 to 6 percent in most of the samples we analyzed from earlier times to its present-day value of around 2 percent at later times and that this was driven by widespread natural selection against Neanderthal DNA.

A large part of the Neanderthal range was in a region where ice ages caused periodic collapses of the animal and plant populations that Neanderthals depended on, a problem that may not have afflicted modern human ancestors in tropical Africa to the same extent. There is genetic confirmation for smaller Neanderthal than modern human population sizes from the fact that the diversity of their genomes was about four times smaller. A history of small size is problematic for the genetic health of a population, because the fluctuations in mutation frequency that occur every generation are substantial enough to allow some mutations to spread through the population even in the face of the prevailing wind of natural selection that tends to reduce their frequencies.[42] So in the half million years since Neanderthals and modern humans separated, Neanderthal genomes accumulated mutations that would prove detrimental when later, Neanderthal/modern human interbreeding occurred.

The problematic mutations in the Neanderthal genome form a sharp contrast with more recent mixtures of divergent modern human populations where there is no evidence for such effects. For example, among African Americans, in studies of about thirty thousand people, we have found no evidence for natural selection against African or European ancestry.[43] One explanation for this is that when Neanderthals and modern humans mixed they had been separated for about ten times longer than had West Africans and Europeans, giving that much more time for biological incompatibilities to develop. A second explanation relates to the observation, from studies of many species, that when infertility arises between populations, it is often due to interactions between two genes in different parts of the genome. Since two changes are required to produce such an incompatibility, the rate of infertility increases with the square of population separation time, so a ten-times-larger population separation translates to one hundred times more genetic incompatibility. In light of this the lack of infertility in hybrids of present-day humans may no longer seem so surprising.

Thesis, Antithesis, Synthesis

An important strand in continental European philosophy beginning in the eighteenth century was that the march of ideas proceeds in a "dialectic": a clash of opposed perspectives that leads to a synthesis.[44] The dialectic begins with a "thesis," followed by an "antithesis." Progress is achieved through a resolution, or "synthesis," which transcends the two-sided debate that engendered it.

So it has been with our understanding of modern human origins. For a long time, many anthropologists favored multiregionalism, the theory that modern humans in any given place in the world descend substantially from archaic humans who lived in the same geographical region. Thus Europeans were thought to derive large proportions of their ancestry from Neanderthals, East Asians from humans who dispersed to eastern Eurasia more than a million years ago, and Africans from African archaic forms. The biological differences among modern human populations would then have extremely deep roots.

Multiregionalism soon encountered its antithesis, the out-of-Africa theory. In this theory, modern humans did not evolve in each location in the world separately from local archaic forms. Instead, modern humans everywhere derive from a relatively recent migration from Africa and the Near East beginning around fifty thousand years ago. The recent date of "Mitochondrial Eve" compared with the deep divergence of Neanderthal mitochondrial DNA provided some of the best evidence for this theory. In opposition to the multiregional hypothesis, the out-of-Africa theory emphasizes the recent origin of the differences among present-day human populations, relative to the multimillion-year time depth of the human skeletal record.

Yet the out-of-Africa argument is not entirely right either. We now have a synthesis, driven by the finding of gene flow between Neanderthals and modern humans based on ancient DNA. This affirms a "mostly out-of-Africa" theory, and also reveals something profound about the culture of those modern humans who must have known Neanderthals intimately. While it is clear from the genetic data that modern humans outside of Africa descend from the expansion of an African-origin group that swept around the world, we now know that some interbreeding occurred. This must make us think differently about our ancestors and the archaic humans they encountered. The Neanderthals were more like us than we had imagined, perhaps capable of many behaviors that we typically associate with modern humans. There must have been cultural exchange that accompanied the mixture—the novels by William Golding and Jean Auel were right to dramatize these encounters. We also know that there has been a biological legacy bequeathed by Neanderthals to non-Africans, including genes for adapting to different Eurasian environments, a topic to which I will return in the next chapter.

At the conclusion of the Neanderthal genome project, I am still amazed by the surprises we encountered. Having found the first evidence of interbreeding between Neanderthals and modern humans, I continue to have nightmares that the finding is some kind of mistake. But the data are sternly consistent: the evidence for Neanderthal interbreeding turns out to be everywhere. As we continue to do genetic work, we keep encountering more and more patterns that reflect the extraordinary impact this interbreeding has had on the genomes of people living today.

So the genetic record has forced our hand. Instead of confirming scientists' expectations, it has produced surprises. We now know that Neanderthal/modern human hybrid populations were living in Europe and across Eurasia, and that while many hybrid populations eventually died out, some survived and gave rise to large numbers of people today. We now know approximately when the modern human and Neanderthal lineages separated. We now also know that when these lineages reencountered each other, they had evolved to such an extent that they were at the very limit of biological compatibility. This raises a question: Were the Neanderthals the only archaic humans who interbred with our ancestors? Or were there other major hybridizations in our past?

A Multiplicity of Archaic Human Lineages

1,400,000–900,000 ya
Main ancestral population of modern humans, Neanderthals, and Denisovans separates from the superarchaic lineage.

770,000–550,000 ya
Genetic estimate of population separation between Neanderthals and modern humans

700,000–50,000 ya
The "Hobbits" persist on the island of Flores in Indonesia.

1,500,000 years ago

PERIOD OF DETAIL

1,000,000–800,000 ya
The Denisovan and Sima de los Huesos mitochondrial DNA lineages separate from those of Neanderthals and modern humans.

470,000–380,000 ya
Genetic estimate of Neanderthal-Denisovan population split

~430,000 ya
Sima de los Huesos skeletons and DNA show that the Neanderthal lineage was already evolving in Europe.

54,000–49,000 ya
Neanderthals and modern humans interbreed.

200,000 ya   100,000   Present

400,000–270,000 ya
Siberian Denisovan and Australo-Denisovan lineages separate.

49,000–44,000 ya
Denisovans and modern humans interbreed.

470,000–360,000 ya
Estimated date at which Neanderthal mitochondrial DNA lineage separates from that of modern humans

500,000 years ago – present

Ancient DNA Opens the Floodgates

A Surprise from the East

In 2008, Russian archaeologists dug up a pinky bone at Denisova Cave in the Altai Mountains of southern Siberia, named after an eighteenth-century Russian hermit named Denis who had made his home there. The bone's growth plates were not fused, showing that the bone came from a child. Its date was uncertain, as it was too small to be dated by radiocarbon analysis, and it was found in a mixed-up soil layer of the cave that contained artifacts dating to both less than thirty thousand and more than fifty thousand years ago. The leader of the excavation, Anatoly Derevianko, reasoned that the bone's owner could have been a modern human, and the sample was so labeled. Alternatively, could the bone's owner have been a Neanderthal, as Neanderthal remains were also found near the cave?[1] Derevianko sent part of the bone to Svante Pääbo in Germany.

Pääbo's team, led by Johannes Krause, was successful in extracting mitochondrial DNA from the Denisova Cave bone.[2] Its sequence was of a type that had never before been observed in more than ten thousand modern human and seven Neanderthal sequences. There are around two hundred mutational differences separating the mitochondrial DNA of people living today from that of Neanderthals. The new mitochondrial DNA from the Denisova finger bone featured nearly four hundred differences from the mitochondrial DNA of both present-day humans and Neanderthals. Based on the rate at which mutations accumulate, mitochondrial DNA sequences from present-day humans and Neanderthals are estimated to have separated from each other 470,000 to 360,000 years ago.[3] The number of mutational differences found in the mitochondrial DNA from the Denisova finger bone suggested a separation time of roughly eight hundred thousand to one million years ago. This suggested that the finger bone might belong to a member of a never-before-sampled group of archaic humans.[4]

The identity of the population, however, was unclear. No skeletons or toolmaking styles existed to give a hint, as they had in the case of the Neanderthals. For Neanderthals, archaeological discoveries had motivated the sequencing of a genome. For this new archaic group, the genetic data came first.

A Genome in Search of a Fossil

I first found out about this previously unknown archaic human population in early 2010, while visiting Svante Pääbo's laboratory in Leipzig, Germany. I was there on one of the thrice-yearly trips I had been making since joining the consortium that Pääbo put together in 2007 to analyze the Neanderthal genome. One evening, Pääbo took me out to a beer garden and told me about the new mitochondrial sequence they had come across. Miraculously, the Denisova finger bone had provided one of the best-preserved samples of ancient DNA ever found. While Pääbo had screened dozens of Neanderthal samples to find a few with up to 4 percent primate DNA, this finger bone had about 70 percent. Pääbo and his team had already been able to obtain more data on the whole genome (not just mitochondrial DNA) from this small bone than they had previously obtained from Neanderthals. He asked if I'd be interested in helping to analyze the data. The invitation to analyze the Denisovan genome was the greatest piece of good fortune I have had in my scientific career.

The mitochondrial genome suggested that the Denisova finger bone came from an individual who was part of a human population that split from the ancestors of modern humans and Neanderthals before they separated from each other. But mitochondrial DNA only records information on the entirely female line, a tiny fraction of the many tens of thousands of lineages that have contributed to any person's genome. To understand what really happened in an individual's history, it is incomparably more valuable to examine all ancestral lineages together. For the Denisova finger bone, the whole genome painted a very different picture from what was recorded in the mitochondrial DNA.

The first revelation from the whole genome was that Neanderthals and the new humans from Denisova Cave were more closely related to each other than either was to modern humans—a different pattern from what was observed in mitochondrial DNA.[5] We eventually estimated the separation between the Neanderthal and Denisovan ancestral populations to have occurred 470,000 to 380,000 years ago, and the separation between the common ancestral populations of both of these archaic groups and modern humans to have occurred 770,000 to 550,000 years ago.[6] The different pattern of relatedness for mitochondrial DNA and the consensus of the rest of the genome were not necessarily a contradiction, as the time in the past when two individuals share a common ancestor at any section of their DNA is always at least as old as the time when their ancestors separated into populations, and can sometimes be far older. However, by studying the whole genome we can learn when the populations split, recognizing that the whole genome encompasses a whole multitude of ancestors so that we can search for short segments of the genome with a relatively low density of mutations reflecting a shared ancestor who lived just before the population separation. Our findings meant that the Denisovans were cousins of Neanderthals, but were also very different, having separated from Neanderthal ancestors before many Neanderthal traits appeared in the fossil record.

We had a heated debate about what to call the new population, and decided to use a generic non-Latin name, "Denisovans," after the cave where they were first discovered, in the same way that Neanderthals are named after the Neander Valley in Germany. This decision distressed some of our colleagues, who lobbied for a new species name—perhaps *Homo altaiensis*, after the mountains where Denisova Cave is located. *Homo altaiensis* is now used in a museum exhibit in Novosibirsk in Russia that describes the discovery at Denisova. We geneticists, however, were reluctant to use a species name. There has long been contention as to whether Neanderthals constitute a species separate from modern humans, with some experts designating Neanderthals as a distinct species of the genus *Homo* (*Homo neanderthalensis*), and others as a subgroup of modern humans (*Homo sapiens neanderthalensis*). The designation of two living groups as distinct species is often based on the supposition that the two do not in practice interbreed.[7] But we now know Neanderthals interbred successfully with modern humans and in fact did so on multiple occasions seems to undermine the argument that they are distinct species. Our data showed that Denisovans were cousins of Neanderthals, and thus if we are uncertain about whether Neanderthals are a species, we need to be uncertain about whether Denisovans are a species as well. Decisions about whether extinct populations are distinct enough to merit designation as different species are traditionally made based on the shapes of skeletons, and for Denisovans there are very few physical remains, providing even more reason to be cautious.

The few remains that we do have are intriguing. Derevianko and his colleagues sent Pääbo a couple of molar teeth from Denisova Cave that contained mitochondrial DNA closely related to the finger bone. These teeth were enormous, beyond the range of nearly all teeth previously reported in the genus *Homo*. Large molars are thought to be biological adaptations to a diet that includes lots of tough uncooked plants. Prior to the Denisovans, the humans closest to us who were known to have had teeth of this size were the primarily plant-eating *australopithecines*, like the famous "Lucy," whose skeleton, dating to more than three million years ago, was found in the Awash Valley of Ethiopia. "Lucy" did not use tools and had a brain only slightly larger than chimpanzees' after correcting for her smaller body size, but she walked upright. Thus the little skeletal information we had confirmed the idea that Denisovans were very distinctive compared to both Neanderthals and modern humans.

The Hybridization Principle

Armed with a whole-genome sequence, we tested whether the Denisovans were more closely related to some present-day populations than others. This led to a huge surprise.

Denisovans were genetically a little closer to New Guineans than they were to any population from mainland Eurasia, suggesting that New Guinean ancestors had interbred with Denisovans. Yet the distance from Denisova Cave to New Guinea is around nine thousand kilometers, and New Guinea is, of course, separated by sea from the Asian mainland. The climate in New Guinea is also largely tropical, which could not be more different from Siberia's bitter winters, and this makes it unlikely that archaic humans adapted to one environment would have flourished in the other.

Skeptical of our findings, we cast around for alternative explanations. Had the ancestors of modern humans been divided into several populations hundreds of thousands of years ago, one of which was more closely related to Denisovans and contributed more to New Guineans' ancestry than it contributed to the ancestry of most other present-day populations? However, this scenario would suggest that the genetic affinity to Denisovans in present-day New Guineans would be due to segments of DNA that entered the New Guinean lineage many hundreds of thousands of years ago. In New Guinean genomes today we were able to measure the size of intact archaic ancestry segments, and found that the ones related to Denisovans were about 12 percent longer than the ones related to Neanderthals, implying that the Denisovan-related segments had been introduced that much more recently on average.[8]

As soon as archaic populations mix with modern ones, the DNA segments contributed by archaic humans are chopped up by the process of recombination, spliced together with modern human segments at the rate of one or two splices per chromosome per generation. As discussed in chapter two, the length of Neanderthal ancestry segments corresponds to mixture between fifty-four and forty-nine thousand years ago.[9] Based on how much longer the

Proportions of Neanderthal Ancestry in People Today

Figure 10. Approximate proportions of Neanderthal (left) and Denisovan ancestry (right) in representative present-day human populations as a fraction of the maximum detected in any group today. Today, Denisovan ancestry is concentrated east of Huxley's Line, a deep-sea trench that

Denisovan segments were than the Neanderthal segments in New Guineans, we could conclude that the interbreeding between Denisovan and New Guinean ancestors occurred forty-nine to forty-four thousand years ago.[10]

What percentage of New Guinean genomes today derives from Denisovans? By measuring how much stronger the genetic evidence of archaic ancestry is in New Guineans compared to other non-Africans, we estimated that about 3 to 6 percent of New Guinean ancestry derives from Denisovans. That is above and beyond the approximately 2 percent from Neanderthals. Thus in total, 5 to 8 percent of New Guinean ancestry comes from archaic humans. This is the largest known contribution of archaic humans to any present-day human population.

Proportions of Denisovan Ancestry in People Today

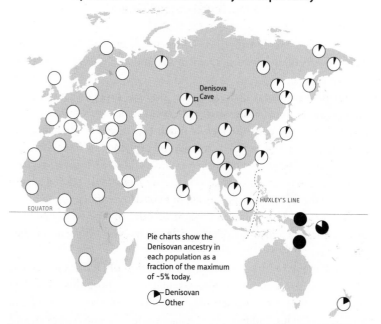

Pie charts show the Denisovan ancestry in each population as a fraction of the maximum of ~5% today.

● Denisovan
○ Other

has always divided mainland Asia from Australia and New Guinea even in the ice ages when sea levels were lower.

The Denisova discovery proved that interbreeding between archaic and modern humans during the migration of modern humans from Africa and the Near East was not a freak event. So far, DNA from two archaic human populations—Neanderthals and Denisovans—has been sequenced, and in both cases, the data made it possible to detect hybridization between modern and archaic humans that had been previously unknown. I would not be surprised if DNA sequenced from the next newly discovered archaic population will also point to a previously unknown hybridization event.

Breaching Huxley's Line

Where, given the vast distance between Siberia and New Guinea, did interbreeding between Denisovans and the ancestors of New Guineans occur?

Our first guess was mainland Asia, perhaps India or central Asia, on a plausible human migratory path from Africa to New Guinea. If this had been the case, the lack of much Denisovan-related ancestry in mainland East or South Asia could be explained by later waves of expansion on the part of modern humans without Denisovan-related ancestry, who replaced populations having Denisovan-related ancestry. That these later migrations did not contribute much to the DNA of present-day New Guineans might account for the relatively high proportion of Denisovan-related ancestry in New Guinean populations today.

A first glance at the geographic distribution of Denisovan-related ancestry in present-day people seemed to support this idea. We collected DNA from present-day humans from the islands of the Southwest Pacific and from East Asia, South Asia, and Australia, and estimated how much Denisovan-related ancestry each of them had. We found the largest amounts of ancestry in indigenous populations in the islands off Southeast Asia and especially in the Philippines and the very large islands of New Guinea and Australia (by the word "indigenous" I refer to people who were established prior to the population movements associated with the spread of farming).[11] The populations in question are largely east of Huxley's Line, a natural boundary that separates New Guinea, Australia, and the Philippines from the western parts of Indonesia and the Asian mainland. This line was described by the nineteenth-century British naturalist Alfred Russel Wallace, and adapted by his contemporary the biologist Thomas Henry Huxley to highlight differences in the animals living on either side, for example, it roughly forms the boundary between placental mammals to the west and marsupials to the east. It corresponds to deep ocean trenches that have formed geographical barriers to the crossing of animals and plants, even in ice ages when sea levels were up to one hundred meters lower. It is remarkable that modern humans after fifty thousand years ago made it across this barrier. These pioneers did manage to cross, but it must have been difficult. Modern humans with Denisovan-related ancestry living east of Huxley's Line—the ancestors of New Guineans, Australians, and Philippine populations who we found are the groups with the largest proportions of Denisovan ancestry today—are likely to have been protected by the same barrier from further migrations from Asia, just like the animals with whom they share their landscape.

But a deeper look suggests that population mixture in the heart of Asia is not as easy an explanation as it might at first seem. Although some populations east of Huxley's Line have large amounts of Denisovan-related ancestry, the situation is very different to the west. Most notably, the indigenous hunter-gatherers of the Andaman Island chain off the coasts of India and Sumatra, and also the indigenous hunter-gatherers of the Malay Peninsula of mainland Southeast Asia, descend from lineages just as divergent as those in indigenous New Guineans and Australians, and yet they do not have much Denisovan-related ancestry. There is also no evidence of elevated Denisovan-related ancestry in genome-wide data from the approximately forty-thousand-year-old human of Tianyuan Cave near Beijing in China, which was sequenced several years later by Pääbo and his laboratory.[12] Had the interbreeding occurred in mainland Asia, and modern humans carrying Denisovan-related ancestry then spread all over, multiple populations of the region as well as ancient humans from East Asia would be expected to carry Denisovan-related ancestry in amounts comparable to what is seen in New Guineans. But this is not what we observe.

The simplest explanation for the large fractions of Denisovan-related ancestry on the islands off the southeastern tip of Asia and in New Guinea and Australia would be the occurrence of interbreeding near the islands—on the islands themselves or in mainland Southeast Asia—but in either case in a tropical region very far from Denisova Cave. However, the anthropologist Yousuke Kaifu pointed out in a talk I attended in 2011 that the hypothesis of interbreeding near the islands is difficult to square with an absence of archaeological artifacts in the region that could plausibly reflect the presence of a big-brained cousin of Neanderthals and modern humans. Kaifu also pointed out that no big-skulled skeletons from this time in this region have so far been found. This makes me think that it is more likely that inter-breeding occurred in southern China or mainland Southeast Asia. There are archaic human remains from Dali in Shaanxi province in north-central China, from Jinniushan in Liaoning in northeastern China, and from Maba in Guangdong in southeastern China, all dating to around two hundred thousand years ago, all of which are more plausible skeletal matches for the Denisovans. An archaic human from Narmada in central India may date to around seventy-five thousand years ago. Chinese and Indian government rules complicate the export of skeletal material, but world-class ancient DNA labs have now been established in China and are beginning to be built in India. DNA from these samples could lead to extraordinary insights.

Meet the Australo-Denisovans

While the interbreeding Neanderthals were close relatives of those we obtained samples from and sequenced, the archaic people who interbred with the ancestors of New Guineans were not close relatives of the Siberian Denisovans. When we examined the genomes of present-day New Guineans and Australians, and counted the number of DNA letter differences between them and the Siberian Denisovans to estimate when their ancestors separated from a common parent population, we discovered that everywhere in the genome, the number of differences was at least what would be expected for a population split that occurred 400,000 to 280,000 years ago.[13] This meant that the ancestors of the Siberian Denisovans separated from the Denisovan lineage that contributed ancestry to New Guineans two-thirds of the way back to the separation of the ancestors of Denisovans from Neanderthals.

In light of the remote relationship, the two groups probably had different adaptations, which would explain how they were able to thrive in such different climates. Given the extraordinary diversity of Denisovans—with much more time separation among their populations than exists among present-day groups—it makes sense to think of them as a broad category of humans, one branch of which became the ancestors of the archaic population that interbred with New Guineans and another that became Siberian Denisovans. Most likely there are other Denisovan populations as well that we haven't sampled at all. Maybe we should even consider Neanderthals as part of this broad Denisovan family.

We never assigned a special name to the Denisovan-related population that interbred with modern humans who migrated to the islands off Southeast Asia, but I like to call them "Australo-Denisovans" to highlight their likely southern geographical distribution. Anthropologist Chris Stringer prefers "Sunda Denisovans" after the landmass that joined most of the Indonesian islands to the Southeast Asian mainland.[14] But this would not be an accurate name if the interbreeding occurred in what is now mainland Southeast Asia, China, or India.

It is tempting to think that the Australo-Denisovans, Denisovans, and Neanderthals descend from the first *Homo erectus* populations that expanded out of Africa, and that modern humans descend from the *Homo erectus* populations that stayed in Africa, but that would be wrong. The oldest *Homo erectus* skeletons outside of Africa have been found at the site of Dmanisi in Georgia dating to around 1.8 million years ago, and on the island of Java in Indonesia dating to almost the same time. If *Homo erectus* from the first radiation out of Africa was ancestral to the Denisovans and Neanderthals, then the split of these populations from modern humans would be at least as old as the dispersal to Eurasia—far too old to be consistent with the genetic observations. The genetic data give a split date of 770,000 to 550,000 years ago, too recent to be consistent with a 1.8-million-year-old population separation.

There is, however, a candidate in the fossil record for an ancestor in the right period, dating to long after the *Homo erectus* out-of-Africa migration but before the *Homo sapiens* one. A big-skulled skeleton found near Heidelberg in Germany in 1907 and dated to around six hundred thousand years ago[15] was plausibly from a species that was ancestral to modern humans and Neanderthals,[16] and by implication, Denisovans too. *Homo heidelbergensis* is often viewed as both a West Eurasian and an African species, but not an East Eurasian species. However, the genetic evidence from the Australo-Denisovans shows that the *Homo heidelbergensis* lineage may have been established very anciently in East Eurasia too. One of the profound implications of the Denisovan discovery was that East Eurasia is a central stage of human evolution and not a sideshow as westerners often assume.

So we now have access to genome-wide data from four highly divergent human populations that all likely had big brains, and that were all still living more recently than seventy thousand years ago. These populations are modern humans, Neanderthals, Siberian Denisovans, and Australo-Denisovans. To these we need to add the tiny humans of Flores island in present-day Indonesia—the "hobbits" who likely descend from early *Homo erectus* whose descendants arrived at Flores island before seven hundred thousand years ago and became isolated there by deep waters.[17] These five groups of humans and probably more groups still undiscovered who lived at that time were each separated by hundreds of thousands of years of evolution. This is greater than the separation times of the most distantly related human lineages today—for example, the one highly represented in San hunter-gatherers from southern Africa and everyone else. Seventy thousand years ago, the world was populated by very diverse human forms, and we have genomes from an increasing number of them, allowing us to peer back to a time when humanity was much more variable than it is today.

How Archaic Encounters Helped Modern Humans

What is the biological legacy of the interbreeding between modern humans and Denisovans? The highest proportion of Denisovan-related ancestry in any present-day population is found in New Guineans and Australians and the people to whom they contributed ancestry.[18] However, once we obtained better data and used more sensitive techniques, we found that there is also some Denisovan-related ancestry, albeit far less, even in mainland Asia,[19] and it is from the mainland that we have a clue about its biological effects.

The Denisovan-related ancestry in East Asians is about a twenty-fifth of that seen in New Guineans—it comprises about 0.2 percent of East Asians' genomes, rising to up to 0.3–0.6 percent in parts of South Asia.[20] We have not yet been able to determine if the Denisovan-related ancestry in mainland Asia and the islands off Southeast Asia comes from the same archaic population or from different ones. If the ancestry comes from very different sources, we would be detecting yet another instance of archaic human interbreeding with modern humans. But whatever its origin, the Denisovan interbreeding was biologically significant.

One of the most striking genomic discoveries of the past few years is a mutation in a gene that is active in red blood cells and that allows people who live in high-altitude Tibet to thrive in their oxygen-poor environment. Rasmus Nielsen and colleagues have shown that the segment of DNA on which this mutation occurs matches much more closely to the Siberian Denisovan genome than to DNA from Neanderthals or present-day Africans.[21] This suggests that some Denisovan relatives in mainland Asia may have harbored an adaptation to high altitude, which the ancestors of Tibetans inherited through Denisovan interbreeding. Archaeological evidence shows that the first inhabitants of the Tibetan high plateau began living there seasonally after eleven thousand years ago, and that permanent occupation based on agriculture began around thirty-six hundred years ago.[22] It is likely that the mutation increased rapidly in frequency only after these dates, a prediction that will be possible to test directly through DNA studies of ancient Tibetans.

Interbreeding with Neanderthals helped modern humans to adapt to new environments just as interbreeding with Denisovans did. We and others showed that at genes associated with the biology of keratin proteins, present-day Europeans and East Asians have inherited much more Neanderthal ancestry on average than is the case for most other groups of genes.[23] This suggests that versions of keratin biology genes carried by Neanderthals were preserved in non-Africans by the pressures of natural selection, perhaps because keratin is an essential ingredient of skin and hair, which are important for providing protection from the elements in cold environments such as the ones that modern humans were moving into and to which Neanderthals were already adapted.

Superarchaic Humans

Given that Denisovans and Neanderthals are genetically closer to each other than either is to modern humans, it would be reasonable to expect them to be equidistantly related to present-day populations that have not received genetic input from either of these archaic populations—that is, to sub-Saharan Africans. Yet we found sub-Saharan Africans to be slightly more closely related to Neanderthals than to Denisovans.[24] This must reflect another example of interbreeding we didn't know about. The pattern we observed could only be explained by Denisovan interbreeding with a deeply divergent, still unknown archaic population—one from which Africans and Neanderthals have little or no DNA, and which separated from the common ancestors of modern humans, Neanderthals, and Denisovans well before their separation from each other.

The evidence for an unknown archaic contribution to Denisovans is that at locations in the genome where all Africans share a mutation, the mutation is more often seen in Neanderthals than in Denisovans. Because these are mutations that all Africans carry, we know that they occurred long ago, as it typically takes around a million years or more in humans for a new mutation not under natural selection to spread throughout a population and achieve 100 percent frequency. The only way to explain the fact that Denisovans do not also share these mutations is if the ancestors of the Denisovans interbred with a population that diverged from Denisovans, Neanderthals, and modern humans so long ago that nearly all modern humans carry the new mutation.

By examining mutations that occur at 100 percent frequency in present-day Africans, and measuring the excess rate at which they matched the Neanderthal over the Denisovan genome, we estimated that the unknown archaic population that interbred into Denisovans first split off from the lineage leading to modern humans 1.4 to 0.9 million years ago and that this unknown archaic population contributed at least 3 to 6 percent of Denisovan-related ancestry. The date is shaky, as knowledge of the human mutation rate is poor. However, even with the uncertainty about the mutation rate, we can estimate relative dates reasonably well, and we can be confident that this previously unsampled human population split off at about twice the separation time of Denisovans, Neanderthals, and modern humans. I think of this group as "superarchaic" humans, as they represent a more deeply splitting lineage than Denisovans. They are what I call a "ghost" population, a population we do not have data from in unmixed form, but whose past existence can be detected from its genetic contributions to later people.

Eurasia as a Hothouse of Human Evolution

From a combination of archaeological and genetic data, we can be confident of at least four major population separations involving modern and archaic human lineages over the last two million years.

The skeletal evidence shows that the first important spread of humans to Eurasia occurred at least 1.8 million years ago, bringing *Homo erectus* from Africa. The genetic evidence suggests that a second lineage split from the one leading to modern humans around 1.4 to 0.9 million years ago, giving rise to the superarchaic group that we have evidence of through its mixture with the ancestors of Denisovans and that plausibly contributed the highly divergent Denisovan mitochondrial DNA sequence that shares a common ancestor with both Neanderthals and modern humans in this time frame. Genetics also suggests a third major split 770,000 to 550,000 years ago when the ancestors of modern humans separated from Denisovans and Neanderthals, followed by Denisovans and Neanderthals from each other 470,000 to 380,000 years ago.

These genetic dates depend on estimates of the mutation rate and will change as those estimates become more exact. It is easy to get ensnared in trying to establish neat correlations between genetic dates and the archaeological record, only to have dates shift when a new genetic estimate of the rate of occurrence of new mutations comes along, causing the whole intellectual edifice to come tumbling down. However, the order of these splits and the distinctness of the populations can be determined well from genetics.

The usual assumption is that all four of these splits correspond to ancestral populations in Africa expanding into Eurasia. But does this really have to be the case?

The argument that modern humans radiated from Africa comes from the observation that the most deeply divergent branches among present-day humans are most strongly represented in African hunter-gatherers (such as San from southern Africa and central African Pygmies). The oldest remains of humans with anatomically modern features are also found in Africa and date to up to around three hundred thousand years ago. However, the genetic comparisons of present-day populations that point to an origin in Africa can only probe the population structure that has arisen in the last couple of hundred thousand years, the time frame of the diversification of the ancestors of present-day populations. With ancient DNA data in hand, we are confronted with the observation that of the four deepest human lineages from which we have DNA data, the three most deeply branching ones are represented only in human specimens excavated from Eurasia: the Neanderthals, the Denisovans, and the "superarchaic" population that left traces among the Siberian Denisovans.

Part of the reason we detect the oldest splitting lineages in Eurasians may be what scientists call "ascertainment bias": the fact that almost all ancient DNA work has been done in Eurasia rather than in Africa, and so naturally that is where new lineages have been discovered. Perhaps if we had as many archaic ancient DNA sequences from Africa as we do from Eurasia, we would find lineages there that split from modern humans and Neanderthals even more deeply in time than the superarchaic.

But another possibility suggests itself, which is that the ancestral population of modern humans, Neanderthals, and Denisovans actually lived in Eurasia, descending from the original *Homo erectus* spread out of Africa. In this scenario, there was later migration back from Eurasia to Africa, providing the primary founders of the population that later evolved into modern humans. The attraction of this theory is its economy: it requires one less major population movement between Africa and Eurasia to explain the data. The superarchaic population and the ancestral population of modern humans, Denisovans, and Neanderthals could both have arisen within Eurasia, without requiring two further out-of-Africa migrations, as long

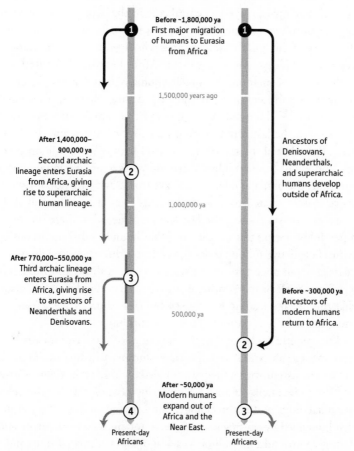

A Plausible Scenario in Which
Modern Human Ancestors Were Not Always in Africa

● Evidence from skeletal remains ○ Evidence from genetic data

Conventional View:
Our Lineage Was Always in Africa.
*At least **four** major migrations*

A Plausible
Alternative Scenario
*As few as **three** major migrations*

Before ~1,800,000 ya
First major migration
of humans to Eurasia
from Africa

1,500,000 years ago

**After 1,400,000–
900,000 ya**
Second archaic
lineage enters Eurasia
from Africa, giving
rise to superarchaic
human lineage.

Ancestors of
Denisovans,
Neanderthals,
and superarchaic
humans develop
outside of Africa.

1,000,000 ya

After 770,000–550,000 ya
Third archaic lineage
enters Eurasia from
Africa, giving rise
to ancestors of
Neanderthals and
Denisovans.

500,000 ya

Before ~300,000 ya
Ancestors of
modern humans
return to Africa.

After ~50,000 ya
Modern humans
expand out of
Africa and the
Near East.

Present-day
Africans

Present-day
Africans

Figure 11. Can the modern human lineage have sojourned for hundreds of thousands of years outside Africa? Conventional models have the human lineage evolving in Africa at all times. To explain current skeletal and genetic data, a minimum of four out-of-Africa migrations are required. However, if our ancestors lived outside Africa from before 1.8 million years ago until up to three hundred thousand years ago, as few as three major migrations would be required.

as there was just one later migration back into Africa to establish shared ancestry with modern humans there.

An argument from economy is not a proof. But the bigger point is that the evidence for many lineages and admixtures should have the effect of shaking our confidence in what to many people is now an unquestioned assumption that Africa has been the epicenter of all major events in human evolution. Based on the skeletal record, it is certain that Africa played a central role in the evolution of our lineage prior to two million years ago, as we have known ever since the discovery of the upright walking apes who lived in Africa millions of years before *Homo*. We know too that Africa has played a central role in the origin of anatomically modern humans, based on the skeletons of humans with anatomically modern features there up to around three hundred thousand years ago, and the genetic evidence for a dispersal in the last fifty thousand years out of Africa and the Near East. But what of the intervening period between two million years ago and about three hundred thousand years ago? In a large part of this time, the human skeletons we have from Africa are not obviously more closely related to modern humans than are the human skeletons of Eurasia.[25] Over the last couple of decades, there has been a pendulum swing toward the view that because our lineage was in Africa before two million years ago and after three hundred thousand years ago, our ancestors must always have been there. But Eurasia is a rich and varied supercontinent, and there is no fundamental reason that the lineage leading to modern humans cannot have sojourned there for an important period before returning to Africa.

The genetic evidence that the ancestors of modern humans may have spent a substantial part of their evolutionary history in Eurasia is in fact consistent with a theory advanced by María Martinón-Torres and Robin Dennell.[26] Theirs is a minority viewpoint within the fields of archaeology and anthropology, but a respected one. They argue that humans they call *Homo antecessor*, found in Atapuerca, Spain, and dating to around one million years ago, show a mix of traits indicating that they are from a population ancestral to modern humans and Neanderthals. This is a very ancient date for a modern human/Neanderthal ancestral population to exist in Eurasia. Many who think that Neanderthals in Europe descend from an out-of-Africa radiation of an ancestral population would assume that the ancestors of both populations were still in Africa at that time. Combining this evidence with archaeological analysis of stone tool types, Martinón-Torres and Dennell argue for the possibility of continuous Eurasian habitation from at least 1.4 million years ago until the most recent common ancestor of humans and Neanderthals after eight hundred thousand years ago, at which point one lineage migrated back to Africa to become the lineage that evolved into modern humans.[27] The Martinón-Torres and Dennell theory becomes more plausible in light of the new genetic evidence.

Part of the "out of Africa" allure is the simplifying idea that Africa—and especially East Africa—has always been the cradle of human diversity and the place where innovation occurred, and that the rest of the world is an evolutionarily inert receptacle. But is there really such a strong case that all the key events in human evolution happened in the same region of the world? The genetic data show that many groups of archaic humans populated Eurasia and that some of these interbred with modern humans. This forces us to question why the direction of migration would have always been out of Africa and into Eurasia, and whether it could sometimes have been the other way around.

The Most Ancient DNA Yet

At the beginning of 2014, Matthias Meyer, Svante Pääbo, and their colleagues in Leipzig extended by a factor of around four the record for the oldest human DNA obtained, sequencing mitochondrial DNA from a more than four-hundred-thousand-year-old *Homo heidelbergensis* individual from the Sima de los Huesos cave system in Spain where twenty-eight ancient humans were found at the bottom of a thirteen-meter shaft.[28] The Sima skeletons have early Neanderthal-like traits, and the archaeologists who excavated them have interpreted them as being on the lineage leading to Neanderthals after the separation from the ancestors of modern humans. Two years after Meyer and Pääbo published mitochondrial DNA data from Sima de los Huesos, they published genome-wide data.[29] Their analysis not only confirmed that the Sima humans were on the Neanderthal lineage, but went further in showing that the Sima humans were more closely related to Neanderthals than they are to Denisovans. These results provided direct evidence that Neanderthal ancestors were already evolving in Europe at least four hundred thousand years ago, and that the separation of the Neanderthal and Denisovan lineages had already begun by that time.

But the Sima data were also perplexing: Sima's mitochondrial genome was more closely related to Denisovans than to Neanderthals, at odds with the genome-wide pattern of it being most closely related to Neanderthals.[30] If there were only one discrepancy between the average relationship measured by the whole genome and the relationship seen in mitochondrial DNA, it might just be possible to believe that this was a statistical fluctuation. But there are two discrepancies in the genetic relationships: the fact that the Sima de los Huesos individual has Denisovan-type mitochondrial DNA despite being closer to Neanderthals in the rest of the genome, and the fact that the Siberian Denisovan individual has mitochondrial DNA twice as divergent from modern humans and Neanderthals as they were from each other despite being closer to Neanderthals in the rest of the genome.[31] The coincidence of these two observations is so improbable that it seems more likely that there is a deeper story to unravel.

Perhaps the superarchaic humans—the ones who interbred with Denisovans—were a much more important part of Eurasian human population history than we initially imagined. Maybe, after separating from the lineage leading to modern humans around 1.4 to 0.9 million years ago, these superarchaic humans spread across Eurasia and began to evolve the ancient mitochondrial lineage found in the Denisovans and Sima humans. At roughly half this time, another group may have split off the lineage leading to modern humans and then spread throughout Eurasia. This group may have mixed into the superarchaic population, contributing the largest proportion of ancestry to populations in the west that evolved into Neanderthals, and a smaller but still substantial proportion of ancestry to populations in the east that became the ancestors of Denisovans. This scenario would explain the findings of two anciently divergent mito-chondrial DNA types in the different groups. It could also explain an odd unpublished observation I have: that in studying the variation in the time since the common genetic ancestor of modern human genomes with both Denisovan and Neanderthal genomes, I have not been able to find evidence for a superarchaic population that con-tributed to Denisovans but not to Neanderthals. Instead the patterns suggest that Denisovans and Neanderthals both had ancestry from the same superarchaic population, with just a larger proportion pres-ent in the Denisovans.

Johannes Krause and colleagues have suggested an alternative the-ory. Krause's idea is that several hundred thousand years ago, an early modern human population migrated out of Africa and mixed with groups like the one that lived in Sima de los Huesos, replacing their mitochondrial DNA along with a bit of the rest of their genomes and creating a mixed population that evolved into true Neanderthals.[32] The idea might seem complicated, but in fact it could explain mul-tiple disparate observations beyond the fact that Neanderthals had a mitochondrial sequence much more similar to modern humans than it did to either the Sima de los Huesos individual or the Siberian Denisovan. It could account for the fact that the estimated date of the common ancestor of humans and Neanderthals in mitochondrial DNA (470,000 to 360,000 years ago)[33] is paradoxically more recent than the estimated date of separation of the ancestors of these two populations based on the analysis of the whole genome (770,000 to 550,000 years ago).[34] It could also explain how it was that Nean-derthals and modern humans both used complex Middle Stone Age methods of manufacturing stone tools, even though the earliest evi-dence for this tool type is hundreds of thousands of years after the genetically estimated separation of the Neanderthals and modern human lineages.[35] The theory finally becomes more plausible in light of a study led by Sergi Castellano and Adam Siepel that suggested up to 2 percent interbreeding into the ancestors of Neanderthals from an early modern human lineage.[36] If Krause's theory is right, this could have been the lineage that spread the mitochondrial DNA found in all Neanderthals.

Whatever explains these patterns, it is clear that we have much more to learn. The period before fifty thousand years ago was a busy time in Eurasia, with multiple human populations arriving from Africa beginning at least 1.8 million years ago. These populations split into sister groups, diverged, and mixed again with each other and with new arrivals. Most of those groups have since gone extinct, at least in their "pure" forms. We have known for a while, from skeletons and archaeology, that there was some impressive human diversity prior to the migration of modern humans out of Africa. However, we did not know before ancient DNA was extracted and studied that Eurasia was a locus of human evolution that rivaled Africa. Against this background, the fierce debates about whether modern humans and Neanderthals interbred when they met in western Eurasia—which have been definitively resolved in favor of interbreeding events that made a contribution to billions of people living today—seem merely anticipatory. Europe is a peninsula, a modest-sized tip of Eurasia. Given the wide diversity of Denisovans and Neanderthals—already represented in DNA sequences from at least three populations separated from each other by hundreds of thousands of years, namely Siberian Denisovans, Australo-Denisovans, and Neanderthals—the right way to view these populations is as members of a loosely related family of highly evolved archaic humans who inhabited a vast region of Eurasia.

Ancient DNA has allowed us to peer deep into time, and forced us to question our understanding of the past. If the first Neanderthal genome published in 2010 opened a sluice in the dam of knowledge about the deep past, the Denisova genome and subsequent ancient DNA discoveries opened the floodgates, producing a torrent of findings that have disrupted many of the comfortable understandings we had before. And that was only the beginning.

Part II

How We Got to Where We Are Today

Modern Humans in Eurasia

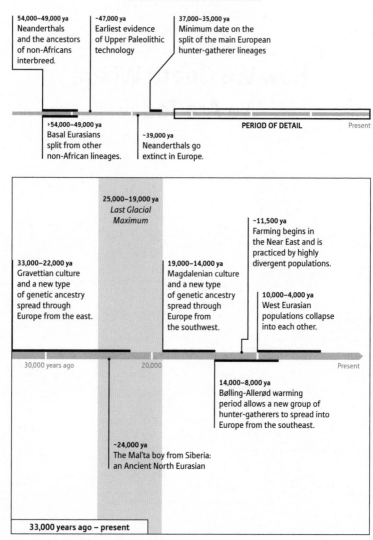

54,000–49,000 ya
Neanderthals and the ancestors of non-Africans interbreed.

~47,000 ya
Earliest evidence of Upper Paleolithic technology

37,000–35,000 ya
Minimum date on the split of the main European hunter-gatherer lineages

›54,000–49,000 ya
Basal Eurasians split from other non-African lineages.

~39,000 ya
Neanderthals go extinct in Europe.

PERIOD OF DETAIL

Present

25,000–19,000 ya
Last Glacial Maximum

~11,500 ya
Farming begins in the Near East and is practiced by highly divergent populations.

33,000–22,000 ya
Gravettian culture and a new type of genetic ancestry spread through Europe from the east.

19,000–14,000 ya
Magdalenian culture and a new type of genetic ancestry spread through Europe from the southwest.

10,000–4,000 ya
West Eurasian populations collapse into each other.

30,000 years ago

20,000

Present

14,000–8,000 ya
Bølling-Allerød warming period allows a new group of hunter-gatherers to spread into Europe from the southeast.

~24,000 ya
The Mal'ta boy from Siberia: an Ancient North Eurasian

33,000 years ago – present

4

Humanity's Ghosts

The Discovery of the Ancient North Eurasians

When confronted with the diversity of life, evolutionary biologists are drawn to the metaphor of a tree. Charles Darwin, at the inception of the field, wrote: "The affinities of all the beings of the same class have sometimes been represented by a great tree. . . . The green and budding twigs may represent existing species. . . . The limbs divided into great branches, and these into lesser and lesser branches, were themselves once, when the tree was small, budding twigs."[1] Present populations budded from past ones, which branched from a common root in Africa. If the tree metaphor is right, then any population today will have a single ancestral population at each point in the past. The significance of the tree is that once a population separates, it does not remix, as fusions of branches cannot occur.

The avalanche of new data that has become available in the wake of the genome revolution has shown just how wrong the tree metaphor is for summarizing the relationship among modern human populations. My closest collaborator, the applied mathematician Nick Patterson, developed a series of formal tests to evaluate whether a tree model is an accurate summary of real population relationships. Foremost among these was the Four Population Test, which, as described in part I, examines hundreds of thousands of positions on the genome where individuals vary—for example, where some people have an adenine (one of the four nucleic acids or "letters" of DNA) and others have a guanine—reflecting a mutation that occurred deep in the past. If a set of four populations is described by a tree, then the frequencies of their mutations are expected to have a simple relationship.[2]

The most natural way to test the tree model is to measure the frequencies of mutations in the genomes of two populations that we hypothesize have split from the same branch. If a tree model is correct, the frequencies of mutations in the two populations will have changed randomly since their separation from the other two more distantly related populations, and so the frequency differences between these two pairs of populations will be statistically independent. If a tree model is wrong, there will be a correlation between the frequency differences, pointing to the likelihood of mixture between the branches. The Four Population Test was central to our demonstration that Neanderthals are more closely related to non-Africans than to Africans, and thus that there was interbreeding between Neanderthals and non-Africans.[3] But findings about interbreeding between archaic and modern humans are only a small part of what has been discovered with Four Population Tests.

My laboratory's first major discovery using the Four Population Test came when we tested the widely held view that Native Americans and East Asians are "sister populations" that descend from a common ancestral branch that separated earlier from the ancestors of Europeans and sub-Saharan Africans. To our surprise, we found that at mutations not shared with sub-Saharan Africans, Europeans are more closely related to Native Americans than they are to East Asians. It would be tempting to argue that this observation has a trivial explanation, such as Native Americans having some ancestry from European migrants over the last five hundred years. But we found the same pattern in every Native American population we studied, including those we could prove had no European admixture. The scenario of Native Americans and Europeans descending from a common population that split earlier from East Asians was also contradicted by the data. Something was deeply wrong with the standard tree model of population relationships.

We wrote a paper describing these results, suggesting that the patterns reflect an episode of mixture deep in the ancestry of Native Americans: a coming together of people related to Europeans and people related to East Asians prior to crossing the Bering land bridge between Asia and the Americas. We submitted this paper, "Ancient Mixture in the Ancestry of Native Americans," in 2009. It was accepted pending minor revisions, but as it turns out, we never published it.

Even as we were making our final revisions to that paper, Patterson discovered something even stranger, which made us realize we had understood only part of the story.[4] To explain his discovery, I need to describe another statistical test we devised, the Three Population Test, which evaluates a "test" population for evidence of mixture. If the test population is a mixture of lineages related to the comparison populations in two different ways—as African Americans are a mixture of Europeans and West Africans—then the frequencies of the test population's mutations are expected to be intermediate between those of the two comparison populations. In contrast, if mixture did not occur, there is no reason to expect the frequencies of mutations in the population to be intermediate. Thus the scenarios of mixture and no mixture yield two qualitatively very different patterns.

When we applied the Three Population Test to diverse human populations, we detected negative statistics when the test population was northern European, proving that population mixture occurred in the ancestors of northern Europeans. We tried all possible pairs of comparison populations from more than fifty worldwide populations and found that the mixture evidence was strongest when one comparison population was southern European, especially Sardinians, and the other was Native Americans. It was clearly Native American populations that produced the most negative values, as we found that the statistic was more negative when we used Native Americans for the second comparison population than when we used East Asians, Siberians, or New Guineans. What we had found was evidence that people in northern Europe, such as the French, are descended from a mixture of populations, one of which shared more ancestry with present-day Native Americans than with any other population living today.

How could we understand the results of both the Three Population Test and the Four Population Test? We proposed that more than fifteen thousand years ago, there was a population living in

Figure 12

Finding the Ghost of North Eurasia

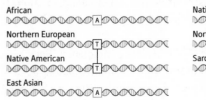

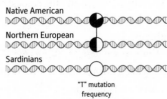

African

Northern European

Native American

East Asian

Native American

Northern European

Sardinians

"T" mutation
frequency

❶ The Four Population Test shows that either northern Europeans have Native American–related mixture or Native Americans have European-related mixture.

❷ The Three Population Test shows that northern Europeans have Native American–related mixture by revealing that mutation frequencies in northern Europeans tend to be intermediate between those in Native Americans and southern Europeans.

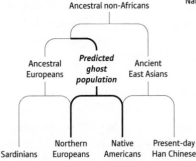

Ancestral non-Africans

Ancestral Europeans

Predicted ghost population

Ancient East Asians

Sardinians

Northern Europeans

Native Americans

Present-day Han Chinese

❸ The existence of Ancient North Eurasians—a population that must have existed in the past and that admixed into both northern Europeans and Native Americans—would explain the test results.

❹ The ghost is found: The genome of the Mal'ta boy from ~24,000 years ago matches the predicted Ancient North Eurasian population.

northern Eurasia that was not the primary ancestral population of the present-day inhabitants of the region. Some people from this population migrated east across Siberia and contributed to the population that crossed the Bering land bridge and gave rise to Native Americans. Others migrated west and contributed to Europeans. This would explain why today, the evidence of mixture in Europeans is strong when using Native Americans as a surrogate for the ancestral population and not as strong in indigenous Siberians, who plausibly descend from more recent, post–ice age migrations into Siberia from more southern parts of East Asia.

We called this proposed new population the "Ancient North Eurasians." At the time we proposed them, they were a "ghost"—a population that we can infer existed in the past based on statistical reconstruction but that no longer exists in unmixed form. The Ancient North Eurasians would without a doubt have been called a "race" had they lived today, as we could show that they must have been genetically about as differentiated from all other Eurasian populations who lived at the time as today's "West Eurasians," "Native Americans," and "East Asians" are from one another. Although they have not left unmixed descendants, the Ancient North Eurasians have in fact been extraordinarily successful. If we put together all the genetic material that they have contributed to present-day populations, they account for literally hundreds of millions of genomes' worth of people. All told, more than half the world's population derives between 5 percent and 40 percent of their genomes from the Ancient North Eurasians.

The case of the Ancient North Eurasians showed that while a tree is a good analogy for the relationships among species—because species rarely interbreed and so like real tree limbs are not expected to grow back together after they branch[5]—it is a dangerous analogy for human populations. The genome revolution has taught us that great mixtures of highly divergent populations have occurred repeatedly.[6] Instead of a tree, a better metaphor may be a trellis, branching and remixing far back into the past.[7]

The Ghost Is Found

At the end of 2013, Eske Willerslev and his colleagues published genome-wide data from the bones of a boy who had lived at the Mal'ta site in south-central Siberia around twenty-four thousand years ago.[8] The Mal'ta genome had its strongest genetic affinity to Europeans and Native Americans, and far less affinity to the Siberians who live in the region today—just as we had predicted for the ghost population of the Ancient North Eurasians. The Mal'ta genome has now become the prototype sample for the Ancient

North Eurasians. Paleontologists would call it a "type specimen," the individual used in the scientific literature to define a newly discovered group.

With the Mal'ta genome in hand, the other pieces of the puzzle snapped into place. It was no longer necessary to reconstruct from present-day populations what had happened long ago. Instead, with a genome sampled directly from the ghost population, it was possible to understand migrations and population admixtures from tens of thousands of years ago as if we were analyzing recent history. What became possible with the Mal'ta genome is the best example I know of the power of ancient DNA to uncover history that until then could only be dimly perceived from present-day data.

The analysis of the Mal'ta genome made it clear that Native Americans derive about a third of their ancestry from the Ancient North Eurasians, and the remainder from East Asians. It is this major mixture that explains why Europeans are genetically closer to Native Americans than they are to East Asians. Our unpublished manuscript claiming that Native Americans descend from a mixture of East Asian and West Eurasian related lineages had been correct, but it was just not the whole story; the paper was overtaken by events in the fast-moving field of ancient DNA. What Willerslev and colleagues found went far beyond what we had been able to do by relying on only modern populations. The Willerslev team not only proved that Native Americans issued from population mixture—which we had not succeeded in doing as we could not rule out an alternative scenario—but they also showed that the mixture was part of a larger story.

The finding that several of the great populations outside of Africa today are profoundly mixed was at odds with what most scientists expected. Prior to the genome revolution, I, like most others, had assumed that the big genetic clusters of populations we see today reflect the deep splits of the past. But in fact the big clusters today are themselves the result of mixtures of very different populations that existed earlier. We have since detected similar patterns in every population we have analyzed: East Asians, South Asians, West Africans, southern Africans. There was never a single trunk population in the human past. It has been mixtures all the way down.

The Ghost of the Near East

Throughout 2013, Iosif Lazaridis in my laboratory was troubled by a result that could not be understood without ancient DNA.

Lazaridis was trying to understand a peculiar Four Population Test result showing that East Asians, present-day Europeans, and pre-farming European hunter-gatherers from around eight thousand years ago are not related to one another according to the tree model. Instead, his analysis showed that East Asians today are genetically more closely related on average to the ancestors of ancient European hunter-gatherers than they are to the ancestors of present Europeans. Ancient DNA studies prior to his work had already shown that present-day Europeans derive some of their ancestry from migrations of farmers from the Near East, who I had assumed were derived from the same ancestral population as European hunter-gatherers. Lazaridis now realized that the ancestry of the first European farmers was distinct from European hunter-gatherers in some way. Something more complicated was going on.

Lazaridis weighed two alternative explanations. One explanation was that there was mixture between the ancestors of ancient European hunter-gatherers and ancient East Asians, bringing these two populations together genetically. There are no insurmountable geographic barriers between Europe and East Asia, so this was a distinct possibility. The alternative explanation was that early European farmers who contributed much of the DNA to present-day Europeans derived some of their ancestry from a population that split early from the main group that peopled Eurasia. This would render East Asians less similar to present-day Europeans than they are to pre-farming European hunter-gatherers.

Once the genome sequence from Mal'ta became available, Lazaridis instantly solved the problem.[9] With Mal'ta in hand, he carried out Four Population Tests among various sets of four populations. Mal'ta and the pre-farming European hunter-gatherers appeared to descend from a common ancestral population that arose after the separation from East Asians and sub-Saharan Africans. The data were consistent with a simple tree. But when Lazaridis replaced ancient European hunter-gatherers in this statistic with either present-day Europeans or with early European farmers, the tree metaphor could no longer describe the data. Present-day Europeans and Near Easterners are mixed: they carry within them ancestry from a divergent Eurasian lineage that branched from Mal'ta, European hunter-gatherers, and East Asians before those three lineages separated from one another.

Lazaridis called this lineage "Basal Eurasian" to denote its position as the deepest split in the radiation of lineages contributing to non-Africans. The Basal Eurasians were a new ghost population, one as important as the Ancient North Eurasians, measured by the sheer number of descendant genomes they have left behind. The extent of the deviations of the Four Population Test away from the value of zero that would be expected if the populations were related by a simple tree indicates that this ghost population contributed about a quarter of the ancestry of present-day Europeans and Near Easterners. It also contributed comparable proportions of ancestry to Iranians and Indians.

No one has yet collected ancient DNA from the Basal Eurasians. Finding such a sample is at present one of the holy grails in the field of ancient DNA, just as finding the Ancient North Eurasians had been before the Mal'ta discovery. But we know that Basal Eurasians existed. And even without having their ancient DNA, we know important facts about them based on the genomic fragments they have left behind in samples for which we do have data.

An extraordinary feature of the Basal Eurasians compared to all other lineages that have contributed to present-day people outside of Africa is that they harbored little or no Neanderthal ancestry. In 2016, we analyzed ancient DNA from the Near East to show that people who lived in the region fourteen thousand to ten thousand years ago had approximately 50 percent Basal Eurasian ancestry, about twice the proportion in Europeans today. Plotting the proportion of Basal Eurasian ancestry against the proportion of Neanderthal ancestry, we realized that the less Basal Eurasian ancestry a non-African person has, the more Neanderthal ancestry he or she has. Thus non-Africans who have zero percent Basal Eurasian ancestry have twice as much Neanderthal DNA as ones with 50 percent Basal Eurasian ancestry. By extrapolation, we might expect 100 percent Basal Eurasians to have no Neanderthal ancestry at all.[10] So wherever the Neanderthal admixture occurred, it seems to have largely happened after the other branches of the non-African family tree separated from Basal Eurasians.

A tempting idea is that the Basal Eurasians represent the descendants of a second wave of migration of modern humans north of the Sahara Desert, well after the dispersal of the population that interbred with Neanderthals. However, this is not correct, as the Basal Eurasian lineage shares much of the history of other non-Africans, including descent from the same relatively small population that founded all non-African lineages more than fifty thousand years ago. The ancient presence of the Basal Eurasians in Eurasia becomes even clearer when one considers that peoples who lived ten thousand years ago or more in what are now Iran and Israel each had around 50 percent Basal Eurasian ancestry,[11] despite the clear genetic evidence that these two populations had been isolated from one another for tens of thousands of years.[12] This suggests the possibility that there were multiple highly divergent Basal Eurasian lineages coexisting in the ancient Near East, not exchanging many migrants until farming expanded. The Basal Eurasians were a major and distinctive source of human genetic variation, with multiple subpopulations persisting for a long period of time.

Where could the Basal Eurasians have lived, isolated as they seem to have been for tens of thousands of years from other non-African lineages? In the absence of ancient DNA, we can only speculate. It is possible that they may have sojourned in North Africa, which is difficult to reach from southern parts of the African continent because of the barrier of the Sahara Desert, and which is more ecologically linked to West Eurasia. Today, the peoples of North Africa owe most of their ancestry to West Eurasian migrants, making the deep genetic past in that region difficult to discern.[13] However, archaeological studies have revealed ancient cultures there that could potentially have corresponded to the Basal Eurasians. The Nile Valley, for example, has been occupied by humans for the entire period since present-day Eurasians diverged from their closest relatives in sub-Saharan Africa.

A hint about the possible homeland of the Basal Eurasians comes from the Natufians, hunter-gatherers who lived after around fourteen thousand years ago in the southwestern parts of the Near East.[14] They were the first people known to have lived in permanent dwellings—they did not migrate from place to place searching for food despite being hunter-gatherers. They built large stone structures and actively managed local wild plants before their successors became full-fledged farmers. Their skulls as well as the stone tools they made are similar in shape to those of North Africans who lived around the same time, and it has been suggested on this basis that the Natufians migrated to the Near East from North Africa.[15] In 2016, my laboratory published ancient DNA from six Natufians from Israel, and we found that they share with early Iranian hunter-gatherers the highest proportions of Basal Eurasian ancestry in the Near East.[16] However, our ancient DNA data cannot determine where the ancestors of the Natufians lived, as we do not yet have comparable ancient DNA data from any other populations that lived at this time or earlier in North Africa, Arabia, or the southwestern Near East. And even if a genetic connection between Natufians and North Africa is established, it will not be the whole story, as it cannot explain the equally high proportions of Basal Eurasian ancestry in the ancient hunter-gatherers and farmers of Iran and the Caucasus.

The Ghosts of Early Europeans

The discovery of one major ghost population after another—Ancient North Eurasians and Basal Eurasians—might make it seem as if ancient DNA is unnecessary, since the existence of ghosts can be predicted from modern populations. But statistical reconstruction can only go so far. With data from present-day people, it is difficult to probe further back in time than the most recent mixture event. Moreover, because humans are so mobile, it is impossible to determine with any confidence where ancestral populations lived based on analyses of the genomes of their descendants. With ancient DNA directly extracted from the ghosts, however, it is possible to project further back in time, revealing even more ancient ghosts than can be recovered from modern data alone. So it was when the Mal'ta genome was sequenced. We discovered the Mal'ta genome statistically, but once we had access to the sequence, we were able to discover the even more distant Basal Eurasians.[17]

In 2016, the lid of Pandora's box opened wide, and a whole mob of ancient ghosts whirled out. My laboratory assembled genome-wide data from fifty-one ancient modern humans in Eurasia, most of them from Europe, who lived between forty-five thousand and seven thousand years ago.[18] These samples spanned the entire period of the Last Glacial Maximum—which occurred between twenty-five thousand and nineteen thousand years ago—when glaciers covered the northern and middle latitudes of Europe so that all humans there lived in refuges in its southern peninsulas. Prior to our work, just a few remains provided genetic data from this period, and the picture that emerged from their analysis was static and monochromatic. But with all our new data, we could show that repeated population transformations, replacements, migrations, and mixtures had taken place over this vast stretch of time.

When analyzing ancient DNA data, the usual approach is to compare ancient individuals to present-day ones, trying to get bearings on the past from the perspective of the present. But when this was done by Qiaomei Fu in my laboratory, her results shed little light on these ancient hunter-gatherers. The differences among humans today are hardly relevant to those that existed in Europe at the time depths she was studying. Fu needed to confront the data on their own terms. To do this, she began by comparing the ancient individuals to one another. She grouped them in four clusters that contained many samples that were similar both genetically and with respect to their archaeologically determined dates. Now she only needed to understand the relationships among the clusters. There were also some individuals who did not cluster with any others, especially among the oldest individuals.

With her samples organized in this way, Fu was able to break down the story of the first thirty-five thousand years of modern humans in West Eurasia into at least five key events.

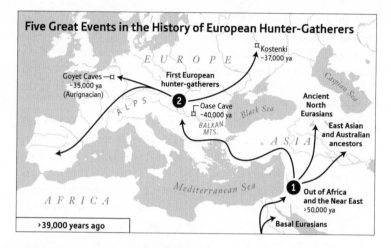

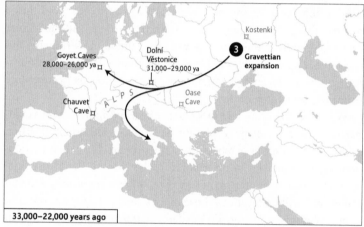

Figure 13. Having migrated out of Africa and the Near East, modern human pioneer populations spread throughout Eurasia (1). By at least thirty-nine thousand years ago, one group founded a lineage of European hunter-gatherers that persisted largely uninterrupted for more than twenty thousand years (2). Eventually, groups derived from an eastern branch of this founding population of European hunter-gatherers spread west (3), displaced previous groups, and were eventually themselves pushed out of northern Europe by the spread of glacial ice, shown at its maximum extent (top right). As the glaciers receded, western Europe was repeopled from the southwest

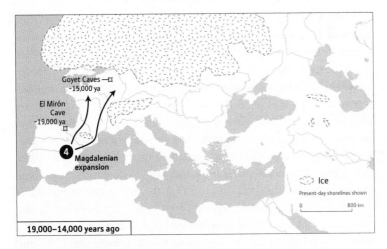

19,000–14,000 years ago

Goyet Caves
~15,000 ya

El Mirón
Cave
~19,000 ya

4 **Magdalenian
expansion**

Ice

Present-day shorelines shown

0   800 km

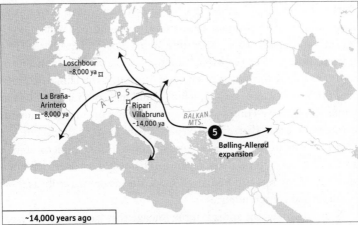

~14,000 years ago

Loschbour
~8,000 ya

La Braña-
Arintero
~8,000 ya

ALPS

Ripari
Villabruna
~14,000 ya

BALKAN
MTS.

5 **Bølling-Allerød
expansion**

(4) by a population that had managed to persist for tens of thousands of years and was related to an approximately thirty-five-thousand-year-old individual from far western Europe. A later human migration, following the first strong warming period, had an even larger impact, with a spread from the southeast (5) that not only transformed the population of western Europe but also homogenized the populations of Europe and the Near East. At a single site—Goyet Caves in Belgium—ancient DNA from individuals spread over twenty thousand years reflects these transformations, with representatives from the Aurignacian, Gravettian, and Magdalenian periods.

Event One was the spread of modern humans into western Eurasia and is evident in the two most ancient samples, an approximately forty-five-thousand-year-old individual whose leg bone had been found eroding out of a riverbank in western Siberia,[19] and an approximately forty-thousand-year-old individual whose lower jaw was found in a cave in Romania.[20] Both individuals were no more closely related to later European hunter-gatherers than they were to present-day East Asians. This finding showed that they were members of pioneer modern human populations that initially flourished but whose descendants largely disappeared. The existence of these pioneer populations makes it clear that the past is not an inevitable march toward the present. Human history is full of dead ends, and we should not expect the people who lived in any one place in the past to be the direct ancestors of those who live there today. Around thirty-nine thousand years ago, a supervolcano near present-day Naples in Italy dropped an estimated three hundred cubic kilometers of ash across Europe, separating archaeological layers preceding it from those that succeeded it.[21] Almost no Neanderthal remains or tools are found above this layer, suggesting that the climate disruption produced by the volcano, which could have produced multiyear winters, may have compounded competition with modern humans to create a crisis that drove Neanderthals to extinction. But the Neanderthals were not the only ones in crisis. Most modern human archaeological cultures that left remains below the ash layer left none above it. Many modern humans disappeared as dramatically as their Neanderthal contemporaries.[22]

Event Two was the spread of the lineage that gave rise to all later hunter-gatherers in Europe. Fu's Four Population Tests showed that both an approximately thirty-seven-thousand-year-old individual from eastern Europe (present-day European Russia)[23] and an approximately thirty-five-thousand-year-old individual from western Europe (present-day Belgium) were part of a population that contributed to all later Europeans, including today's.[24] Fu also used Four Population Tests to show that during the entire period from around thirty-seven thousand to around fourteen thousand years ago, almost all the individuals she analyzed from Europe could be rather well described as descending from a single common ancestral population that had not experienced mixture with non-European populations. Archaeologists have shown that after the volcanic eruption around thirty-nine thousand years ago, a modern human culture spread across Europe making stone tools of a type known as Aurignacian, and that this replaced the diverse stone toolmaking styles that existed before. Thus genetic and archaeological evidence both point to multiple independent migrations of early modern human pioneers into Europe, some of which went extinct and were replaced by a more homogeneous population and culture.

Event Three was the coming of the people who made Gravettian tools, who dominated most of Europe between around thirty-three thousand and twenty-two thousand years ago. The material remains they left behind include voluptuous female statuettes, as well as musical instruments and dazzling cave art. Compared to the people who made Aurignacian tools who came before them, the people who made Gravettian tools were much more deliberate about burying their dead, and as a result we have many more skeletons from this period than we do from the Aurignacian period. We extracted DNA from Gravettian-era individuals buried in present-day Belgium, Italy, France, Germany, and the Czech Republic. They were all genetically very similar despite their extraordinary geographic dispersal. Fu's analysis indicated that most of their ancestry derived from the same sublineage of European hunter-gatherers as the thirty-seven-thousand-year-old individual from far eastern Europe, and that they then spread west, displacing the sublineage associated with Aurignacian tools and represented in the thirty-five-thousand-year-old Belgian individual. The changes in artifact styles associated with the rise of the Gravettian culture were thus driven by the spread of new people.

Event Four was heralded by a skeleton from present-day Spain dating to around nineteen thousand years ago—one of the first individuals known to be associated with the Magdalenian culture, whose members over the next five thousand years migrated to the northeast out of their warm-weather refuge, chasing the retreating ice sheets into present-day France and Germany. The data once again showed a correspondence between the archaeological culture and genetic discoveries, documenting the spread of people into central Europe who were not directly descended from the Gravettians who had preceded them. There was also a surprise: most of the ancestry of individuals associated with the Magdalenian culture came from the sublineage represented by the thirty-five-thousand-year-old individual from Belgium who was associated with Aurignacian tools but who was later succeeded at the same site by people who used Gravettian tools and carried DNA similar to others in Europe associated with that culture of eastern European origin. Here was yet another ghost population that contributed to later groups in mixed form. The Aurignacian lineage had not died out, but instead had persisted in some geographic pocket, possibly in western Europe, before its resurgence at the end of the ice age.

Event Five happened around fourteen thousand years ago, during the first strong warming period after the last ice age, a major climatic change known as the Bølling-Allerød. Geological reconstructions reveal that at this time, the Alpine glacial wall that extended down to the Mediterranean Sea near present-day Nice finally melted after about ten thousand years of dividing the west and east of Europe. Plants and animals from southeastern Europe (the Italian and Balkan peninsulas) migrated in abundance into southwestern Europe.[25] Our Four Population Tests on our ancient DNA data showed that something similar happened with humans. After around fourteen thousand years ago, a group of hunter-gatherers spread across Europe with ancestry quite different from that of the people associated with the preceding Magdalenian culture, whom they largely displaced. Individuals living in Europe between thirty-seven thousand and fourteen thousand years ago were all plausibly descended from a common ancestral population that separated earlier from the ancestors of lineages represented in the Near East today. But after around fourteen thousand years ago, western European hunter-gatherers became much more closely related to present-day Near Easterners. This proved that new migration occurred between the Near East and Europe around this time.

We do not yet have ancient DNA from the period before fourteen thousand years ago from southeastern Europe and the Near East. We can therefore only surmise population movements around this time. The people who had waited out the ice age in southern Europe became dominant across the entire European continent following the melting of the Alpine glacial wall.[26] Perhaps these same people also expanded east into Anatolia, and their descendants spread farther to the Near East, bringing together the genetic heritages of Europe and the Near East more than five thousand years before farmers spread Near Eastern ancestry back into Europe by migrating in the opposite direction.

The Genetic Formation of Present-Day West Eurasians

Today, the peoples of West Eurasia—the vast region spanning Europe, the Near East, and much of central Asia—are genetically highly similar. The physical similarity of West Eurasian populations was recognized in the eighteenth century by scholars who classified the people of West Eurasia as "Caucasoids" to differentiate them from East Asian "Mongoloids," sub-Saharan African "Negroids," and "Australoids" of Australia and New Guinea. In the 2000s, whole-genome data emerged as a more powerful way to cluster present-day human populations than physical features.

The whole-genome data at first seem to validate some of the old categories. The most common way to measure the genetic similarity between two populations is by taking the square of the difference in mutation frequencies between them, and then averaging across thousands of independent mutations across the genome to get a precisely determined number. Measured in this way, populations within West Eurasia are typically around seven times more similar to one another than West Eurasians are to East Asians. When frequencies of mutations are plotted on a map, West Eurasia appears homogeneous, from the Atlantic façade of Europe to the steppes of central Asia. There is a sharp gradient of change in central Asia before another region of homogeneity is reached in East Asia.[27]

How did the present-day population structure emerge from the one that existed in the deep past? We and other ancient DNA laboratories found in 2016 that the formation of the present-day West Eurasian population was propelled by the spread of food producers. Farming began between twelve and eleven thousand years ago in southeastern Turkey and northern Syria, where local hunter-gatherers began domesticating most of the plants and animals many West Eurasians still depend upon today, including wheat, barley, rye, peas, cows, pigs, and sheep. After around nine thousand years ago, farming began spreading west to present-day Greece and roughly at the same time began spreading east, reaching the Indus Valley in present-day Pakistan. Within Europe, farming spread west along the Mediterranean coast to Spain, and northwest to Germany through the Danube River valley, until it reached Scandinavia in the north and the British Isles in the west—the most extreme places where this type of economy was practical.

Until 2016, getting genome-wide ancient DNA from the Near East to assess the extent to which these changes in the archaeological record were propelled by movements of people had failed, as the warm climate of the Near East quickens chemical reactions, accelerating the rate of breakdown of DNA. However, two technical breakthroughs changed this. One came from a method developed by Matthias Meyer, which involved enriching DNA extracted from ancient bones for human sequences of interest.[28] This approach makes ancient DNA analysis up to one thousand times more cost-effective and gives access to samples that would otherwise provide too little DNA to study. Working together with Meyer, we adapted this method to make possible genome-wide analysis of large numbers of samples.[29] The second breakthrough was the recognition that the inner-ear part of the skull—known as the petrous bone—preserves a far higher density of DNA than most other skeletal parts, up to one hundred times more for each milligram of bone powder. Within the petrous bone, the anthropologist Ron Pinhasi, working in Dublin, showed that the mother lode of DNA is found in the cochlea, the snail-shaped organ of hearing.[30] Ancient DNA analysis of petrous bones in 2015 and 2016 broke through one barrier after another and made it possible for the first time to get ancient DNA from the warm Near East.

Working with Pinhasi, we obtained ancient DNA from forty-four ancient Near Easterners across much of the geographic cradle of farming.[31] The results revealed that around ten thousand years ago, at the time that farming was beginning to spread, the population structure of West Eurasia was far from the genetic monoculture we observe today. The farmers of the western mountains of Iran, who may have been the first to domesticate goats, were genetically directly derived from the hunter-gatherers who preceded them. Similarly, the first farmers of present-day Israel and Jordan were descended largely from the Natufian hunter-gatherers who preceded them. But these two populations were also very genetically different from each other. We and another research group[32] found that the degree of genetic differentiation between the first farmers of the western part of the Near East (the Fertile Crescent, including Anatolia and the Levant) and the first farmers of the eastern part (Iran) was about as great as the differentiation between Europeans and East Asians today. In the Near East, the expansion of farming was accomplished not just by the movement of people, as happened in Europe, but also by the spread of common ideas across genetically very different groups.

The high differentiation of human populations in the Near East ten thousand years ago was a specific instance of a broader pattern across the vast region of West Eurasia, documented by Iosif Lazaridis, who led the analysis. Analyzing our data, he found that about ten thousand years ago there were at least four major populations in West Eurasia—the farmers of the Fertile Crescent, the farmers of Iran, the hunter-gatherers of central and western Europe, and the hunter-gatherers of eastern Europe. All these populations differed from one another as much as Europeans differ from East Asians today. Scholars interested in trying to create ancestry-based racial classifications, had they lived ten thousand years ago, would have categorized these groups as "races," even though none of these groups survives in unmixed form today.

Spurred by the revolutionary technology of plant and animal domestication, which could support much higher population densities than hunting and gathering, the farmers of the Near East began migrating and mixing with their neighbors. But instead of one group displacing all the others and pushing them to extinction, as had occurred in some of the previous spreads of hunter-gatherers in Europe, in the Near East all the expanding groups contributed to later populations. The farmers in present-day Turkey expanded

into Europe. The farmers in present-day Israel and Jordan expanded
into East Africa, and their genetic legacy is greatest in present-day
Ethiopia. Farmers related to those in present-day Iran expanded into
India as well as the steppe north of the Black and Caspian seas. They
mixed with local populations there and established new economies
based on herding that allowed the agricultural revolution to spread
into parts of the world inhospitable to domesticated crops. The different food-producing populations also mixed with one another, a
process that was accelerated by technological developments in the
Bronze Age after around five thousand years ago. This meant that
the high genetic substructure that had previously characterized
West Eurasia collapsed into the present-day very low level of genetic
differentiation by the Bronze Age. It is an extraordinary example
of how technology—in this case, domestication—contributed to
homogenization, not just culturally but genetically. It shows that
what is happening with the Industrial Revolution and the information revolution in our own time is not unique in the history of our
species.

The fusion of these highly different populations into today's West
Eurasians is vividly evident in what might be considered the classic northern European look: blue eyes, light skin, and blond hair.
Analysis of ancient DNA data shows that western European hunter-
gatherers around eight thousand years ago had blue eyes but dark
skin and dark hair, a combination that is rare today.[33] The first
farmers of Europe mostly had light skin but dark hair and brown
eyes—thus light skin in Europe largely owes its origins to migrating
farmers.[34] The earliest known example of the classic European blond
hair mutation is in an Ancient North Eurasian from the Lake Baikal
region of eastern Siberia from seventeen thousand years ago.[35] The
hundreds of millions of copies of this mutation in central and western Europe today likely derive from a massive migration into the
region of people bearing Ancient North Eurasian ancestry, an event
that is related in the next chapter.[36]

Surprisingly, the ancient DNA revolution, through its discovery
of the pervasiveness of ghost populations and their mixture, is fuel-
ing a critique of race that has been raised by scholars in the past, but
was never prominent because of a lack of support from hard scien-

tific facts.[37] By demonstrating that the genetic fault lines in West Eurasia between ten thousand and four thousand years ago were entirely different from today's, the ancient DNA revolution has shown that today's classifications do not reflect fundamental "pure" units of biology. Instead, today's divisions are recent phenomena, with their origin in repeating mixtures and migrations. The findings of the ancient DNA revolution suggest that the mixtures will continue. Mixture is fundamental to who we are, and we need to embrace it, not deny that it occurred.

How Europe's Three Ancestral Populations Came Together

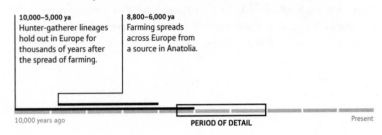

10,000–5,000 ya
Hunter-gatherer lineages hold out in Europe for thousands of years after the spread of farming.

8,800–6,000 ya
Farming spreads across Europe from a source in Anatolia.

10,000 years ago

PERIOD OF DETAIL

Present

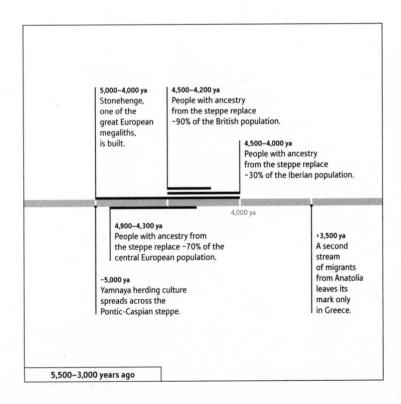

5,000–4,000 ya
Stonehenge, one of the great European megaliths, is built.

4,500–4,200 ya
People with ancestry from the steppe replace ~90% of the British population.

4,500–4,000 ya
People with ancestry from the steppe replace ~30% of the Iberian population.

4,000 ya

4,900–4,300 ya
People with ancestry from the steppe replace ~70% of the central European population.

~5,000 ya
Yamnaya herding culture spreads across the Pontic-Caspian steppe.

›3,500 ya
A second stream of migrants from Anatolia leaves its mark only in Greece.

5,500–3,000 years ago

The Making of Modern Europe

Strange Sardinia

In 2009, geneticists led by Joachim Burger sequenced stretches of mitochondrial DNA from ancient European hunter-gatherers and some of the earliest farmers of Europe.[1] Although mitochondrial DNA is hundreds of thousands of times shorter than the rest of the genome, it has enough variation to allow categorization of the peoples of the world into distinct types. Nearly all ancient hunter-gatherers carried one set of mitochondrial DNA types. But the farmers who succeeded them carried no more than a few percent of those types, and their DNA was more similar to that seen today in southern Europe and the Near East. It was clear that the farmers came from a population that did not descend from European hunter-gatherers.

Mitochondrial DNA is only a small portion of the genome, however, and the whole-genome studies that followed delivered strange results. In 2012, a team of geneticists sequenced the genome of the "Iceman," a natural mummy dating to approximately fifty-three hundred years ago that was discovered in 1991 on a melting glacier in the Alps.[2] The cold had preserved his body and equipment, providing a vivid snapshot of what obviously had been an extraordinarily complex culture dating to thousands of years before the arrival of writing. His skin was covered with dozens of tattoos. He wore a woven grass cloak and finely sewn shoes. He carried a copper-bladed axe

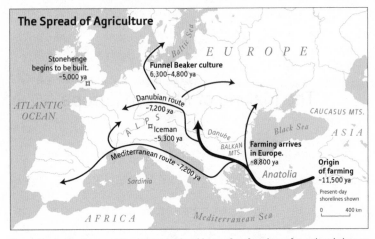

Figure 14a. Archaeology and linguistics provide evidence of profound transformations in human culture. Archaeological evidence shows that farming expanded from the Near East to the far northwest of Europe between about 11,500 years ago and about 5,500 years ago, transforming economies across this region.

and a kit for lighting fires. An arrowhead in his shoulder and a torn artery showed that he had been shot and had stumbled to the top of a mountain pass before collapsing. Based on the isotopes of the elements strontium, lead, and oxygen in the enamel capping his teeth, it seemed likely he had grown up in a nearby valley where isotopes (contained in groundwater and plants, and derived from the local rocks) had similar ratios.[3] But the ancient DNA data showed that his closest genetic relatives are not present-day Alpine people. Instead, his closest relatives today are the people of Sardinia, an island in the Mediterranean Sea.

This strange link to present-day Sardinians kept turning up. In the same year that the Iceman's genome was published, Pontus Skoglund, Mattias Jakobsson, and colleagues at the University of Uppsala published four genome sequences from individuals who lived about five thousand years ago in Sweden.[4] A leading theory up until their study was that the Swedish hunter-gatherers who lived at that time descended from farmers who had adapted a hunter-gatherer lifestyle to exploit the rich fisheries of the Baltic Sea, and

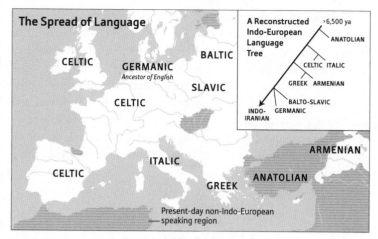

Figure 14b. European languages are nearly all part of the Indo-European language family that descends from a common ancestral language as recently as about 6,500 years ago. (The map labels show the pre-Roman distribution of Indo-European languages.)

were not directly descended from the hunter-gatherers who had lived in northern Europe (including Sweden) several thousand years earlier. But ancient DNA disproved this theory. Instead of being genetically close to each other, the farmers and hunter-gatherers were almost as different from each other as Europeans are from East Asians today. And the farmers once again had that strange link to Sardinians.

Skoglund and Jakobsson proposed a new model to explain these findings—that migrating farmers whose ancestors originated in the Near East spread over Europe with little mixture with the hunter-gatherers they encountered along the way, a sharp contrast to Luca Cavalli-Sforza's model for the farming expansion into Europe that had been popular until this time and that emphasized extensive mixture and interaction with the local hunter-gatherers during the expansion.[5] The new model would not only explain the striking genetic contrast between hunter-gatherers and farmers in Sweden around five thousand years ago. It would also explain why the ancient farmers were genetically similar to present-day Sardinians, who plausibly descend from a migration of farmers to that island around eight thousand years ago that largely displaced the previous hunter-gatherers. Isolated on Sardinia, the descendants of these farmers were minimally affected by demographic events that later transformed the populations of mainland Europe. So far, so good— this new model explained the genetic composition of most Europeans up until around five thousand years ago. But Skoglund and Jakobsson also went further and proposed that these two sources— hunter-gatherers and farmers—might have contributed almost all the ancestry of Europeans living today. Here they missed something extraordinarily important.

A Cloud on the Horizon

In 2012, it seemed that the big question of the ancestral sources of present-day European populations might be solved. But there was an observation that didn't fit.

In that year, Nick Patterson published a perplexing result from his Three Population Test. As described in the previous chapter, he showed that the frequencies of mutations in northern Europeans today tend to be intermediate between those of southern Europeans and Native Americans. He hypothesized that these findings could be explained by the existence of a "ghost population"— the Ancient North Eurasians—who were distributed across northern Eurasia more than fifteen thousand years ago and who contributed both to the population that migrated across the Bering land bridge to people the Americas and to northern Europeans.[6] A year later, Eske Willerslev and colleagues obtained a sample of ancient DNA from Siberia that matched the predicted Ancient North Eurasians— the Mal'ta individual whose skeleton dated to around twenty-four thousand years ago.[7]

How could the finding of an Ancient North Eurasian contribution to present-day northern Europeans be reconciled with the two-way mixture of indigenous European hunter-gatherers and incoming farmers from Anatolia that had been directly demonstrated through ancient DNA studies? The plot became even thicker as we and others obtained additional ancient DNA data from hunter-gatherers and farmers between eight thousand and five thousand years ago and found that they fit the two-way mixture model without any evidence of Ancient North Eurasian ancestry.[8] Something profound must have happened later—a new stream of migrants must have arrived, introducing Ancient North Eurasian ancestry and transforming Europe.

In 2014–15, the ancient DNA community and especially my own laboratory published data from more than two hundred ancient Europeans from Germany, Spain, Hungary, the steppe of far eastern Europe, and the first farmers from Anatolia.[9] By comparing the ancient individuals to West Eurasian people living today, Iosif Lazaridis in my laboratory was able to figure out how it was that the Ancient North Eurasian ancestry entered Europe within the last five thousand years.

Our initial approach was to carry out a principal component analysis, which can identify combinations of mutation frequencies that are most efficient at finding differences among samples. In doing this, we benefited from our extraordinarily high resolution data from around six hundred thousand variable locations on the genome, around ten thousand times more locations than Cavalli-Sforza had been able to analyze in his 1994 book.[10] While Cavalli-Sforza had tried to make sense of the principal component summaries of genetic variation by plotting their values onto a map of the world, we could do far more. We plotted a single dot for each individual depending on where he or she fell relative to the two principal components. On the scatterplot we obtained for close to eight hundred present-day West Eurasians, two parallel lines appeared: the left containing almost all Europeans, and the right containing almost all Near Easterners, with a striking gap in between. By placing all the ancient samples onto the same plot, we could watch their positions shift over time, and the last eight thousand years of European history unfurled before our eyes, offering a time-lapse video showing how present-day Europeans formed from populations that had little resemblance in their ancestry to most Europeans living today.[11]

First came the hunter-gatherers, who themselves were the product of a series of population transformations over the previous thirty-five thousand years as described in the last chapter, the most recent

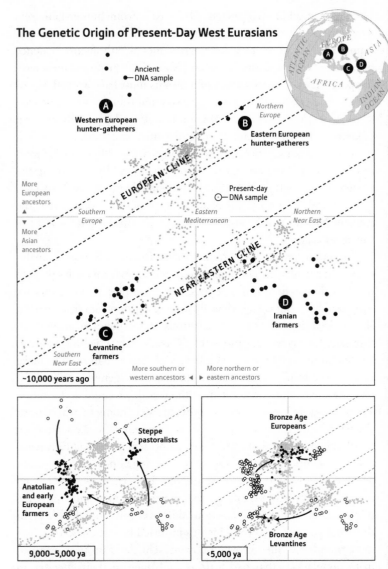

The Genetic Origin of Present-Day West Eurasians

Figure 15. This plot shows a statistical analysis of the primary gradients of genetic variation in present-day people (gray dots) and ancient West Eurasians (black and open dots). Ten thousand years ago, West Eurasia was home to four populations as differentiated from one another as Europeans and East Asians are today. The farmers of Europe and western Anatolia from nine thousand to five thousand years ago were a mixture of western European hunter-gatherers (A), Levantine farmers (C) and Iranian farmers (D). Meanwhile, the pastoralists of the steppe north of the Black and Caspian seas around five thousand years ago were a mixture of eastern European hunter-gatherers (B) and Iranian farmers (D). In the Bronze Age, these mixed populations mixed further to form populations with ancestry similar to people today.

of which was a massive expansion of people out of southeastern Europe by around fourteen thousand years ago that displaced much of the previously established population.[12] In principal component analysis, the hunter-gatherers who lived in Europe at this time fell beyond present-day Europeans along an axis measuring the difference between Europe and the Near East. This was consistent with their having contributed ancestry to present-day Europeans but not to present-day Near Easterners.

Second came the first farmers, who lived between about eighty-eight hundred and forty-five hundred years ago in Germany, Spain, Hungary, and Anatolia. Ancient farmers from all these places were genetically similar to present-day Sardinians, showing that a pioneer farmer population had landed in Greece probably from Anatolia, and then spread to Iberia in the west and Germany in the north, retaining at least 90 percent of their DNA from that immigrant source, which meant that they mixed minimally with the hunter-gatherers they encountered along the way. Further investigation, though, showed that it was not quite so simple. We also found that farmers from the Peloponnese in southern Greece who lived around six thousand years ago may have derived part of their ancestry from a different source population in Anatolia—a population that descended more from Iranian-related populations than was the case in the northwestern Anatolian farmers who were a likely source population for the rest of Europe's farmers.[13] The first farming in Europe was practiced in the Peloponnese and the nearby island of Crete by people who did not use pottery. This has led some archaeologists to wonder if they were from a different migration.[14] Our ancient DNA is consistent with this idea, and suggests the possibility that this population held on for thousands of years.

Third, we identified a new development in farmers living between six thousand and forty-five hundred years ago. In many of these later farmers, we observed a shift toward approximately 20 percent extra hunter-gatherer ancestry, not present in the early farmers, implying that genetic mixing between the previously established people and new arrivals had begun, albeit after a couple of thousand years' delay.[15]

How did the farming and hunter-gatherer cultures coexist? Hints come from the Funnel Beaker culture, which is named for decorated clay vessels in graves dated after about sixty-three hundred years ago. The Funnel Beaker culture arose in a belt of land a few hundred kilometers from the Baltic Sea, which was not reached by the first wave of farmers, probably because their methods were not optimized for the heavy soils of northern Europe. Protected by the stronghold of their difficult-to-farm environment, and sustained by the fish and game resources of Baltic Europe, the northern hunter-gatherers had more than a thousand years to adapt to the challenge of farming. They adopted domesticated animals, and later crops, from their southern neighbors, but kept many elements of their hunter-gathering ways. The people of the Funnel Beaker culture were among those who built megaliths, the collective burial tombs made of stones so large it would have taken dozens of people to move them. The archaeologist Colin Renfrew suggested that megalith building might be a direct reflection of this boundary between southern farmers and hunter-gatherers turned farmers—a way of laying claim to territory, of distinguishing one people and culture from others.[16] The genetic data may bear witness to this interaction, as there was clearly a stream of new migrants into the mixed population. Between six thousand and five thousand years ago, most of the northern gene pool was overtaken by farmer ancestry, and it was this mixture of a modest amount of hunter-gatherer-related ancestry and a large amount of Anatolian farmer–related ancestry—in a population that retained key elements of hunter-gatherer culture—that characterized the Funnel Beaker potters and many other contemporary Europeans.

Europe had reached a new equilibrium. The unmixed hunter-gatherers were disappearing, persisting only in isolated pockets like the islands off southern Sweden. In southeastern Europe, a settled farmer population had developed the most socially stratified societies known up until that time, and rituals that as the archaeologist Marija Gimbutas showed featured women in a central way—a far cry from the male-centered rituals that followed.[17] In remote Britain, the megalith builders were hard at work on what developed into the greatest man-made monument the world had seen: the standing stones of Stonehenge, which became a national place of pilgrimage as reflected by goods brought from the far corners of Britain. People like those at Stonehenge were building great temples to their gods, and tombs for their dead, and could not have known that within a few hundred years their descendants would be gone and their lands overrun. The extraordinary fact that emerges from ancient DNA is that just five thousand years ago, the people who are now the primary ancestors of all extant northern Europeans had not yet arrived.

The Tide from the East

The grasslands of the steppe stretch about eight thousand kilometers from central Europe to China. Prior to five thousand years ago, the archaeological evidence indicates that almost no one lived far from the steppe river valleys, because in between these areas there was too little rain to support agriculture, and too few watering holes to support livestock. The European third of the steppe was a hodgepodge of local cultures, each with its own pottery style, spread thinly over the landscape in places where water could be found.[18]

All this changed with the emergence of the Yamnaya culture around five thousand years ago, whose economy was based on sheep and cattle herding. The Yamnaya emerged from previous cultures of the steppe and its periphery and exploited the steppe resources far more effectively than their predecessors. They spread over a vast region, from Hungary in Europe to the foothills of the Altai Mountains in central Asia, and in many places replaced the disparate cultures that had preceded them with a more homogeneous way of life.

One of the inventions that drove the spread of the Yamnaya was the wheel, whose geographic origin is not known because once it appeared—at least a few hundred years before the rise of the Yamnaya—it spread across Eurasia like wildfire. Wagons using wheels may have been adopted by the Yamnaya from their neighbors to the south: the Maikop culture in the Caucasus region between the Black and Caspian seas. For the Maikop, as for many cultures across Eurasia, the wheel was profoundly important. But for the people of the steppe, it was if anything even more important, as it made possible an economy and culture that were entirely new. By hitching their animals to wagons, the Yamnaya could take water and supplies with them into the open steppe and exploit the vast lands that had previously been inaccessible. By taking advantage of another innovation—the horse, which had recently been domesticated in a more eastern part of the steppe, and which made cattle herding more efficient as a single rider could herd many times the number of animals than could be herded by a person on foot—the Yamnaya also became vastly more productive.[19]

The profound transformation in culture that began with the Yamnaya is obvious to many archaeologists of the steppe. The increase in the intensity of the human use of the steppe lands coincided with a nearly complete disappearance of permanent settlements—almost all the structures that the Yamnaya left behind were graves, huge mounds of earth called kurgans. Sometimes people were buried in kurgans with wagons and horses, highlighting the importance of horses to their lifestyle. The wheel and horse so profoundly altered the economy that they led to the abandonment of village life. People lived on the move, in ancient versions of mobile homes.

Prior to the explosion of ancient DNA data in 2015, most archaeologists found it inconceivable that the genetic changes associated with the spread of the Yamnaya culture could be as dramatic as the archaeological changes. Even the archaeologist David Anthony, a leading proponent of the idea that the spread of Yamnaya culture was transformative in the history of Eurasia, could not bring himself to suggest that its spread was driven by mass migration. Instead, he proposed that most aspects of Yamnaya culture spread through imitation and proselytization.'[20]

But the genetics showed otherwise. Our analysis of DNA from the Yamnaya—led by Iosif Lazaridis in my laboratory—showed that they harbored a combination of ancestries that did not previously exist in central Europe. The Yamnaya were the missing ingredient, carrying exactly the type of ancestry that needed to be added to early European farmers and hunter-gatherers to produce populations with the mixture of ancestries observed in Europe today.[21] Our ancient DNA data also allowed us to learn how the Yamnaya themselves had formed from earlier populations. From seven thousand until five thousand years ago, we observed a steady influx into the steppe of a population whose ancestors traced their origin to the south—as it bore genetic affinity to ancient and present-day people of Armenia and Iran—eventually crystallizing in the Yamnaya, who were about a one-to-one ratio of ancestry from these two sources.[22] A good guess is that the migration proceeded via the Caucasus isthmus between the Black and Caspian seas. Ancient DNA data produced by Wolfgang Haak, Johannes Krause, and their colleagues have shown that the populations of the northern Caucasus had ancestry of this type continuing up until the time of the Maikop culture, which just preceded the Yamnaya.

The evidence that people of the Maikop culture or the people who preceded them in the Caucasus made a genetic contribution to the Yamnaya is not surprising in light of the cultural influence the Maikop had on the Yamnaya. Not only did the Maikop pass on to the Yamnaya their technology of carts, but they were also the first to build the kurgans that characterized the steppe cultures for thousands of years afterward. The penetration of Maikop lands by Iranian- and Armenian-related ancestry from the south is also plausible in light of studies showing that Maikop goods were heavily influenced by elements of the Uruk civilization of Mesopotamia to the south, which was poor in metal resources and engaged in trade and exchange with the north as reflected in Uruk goods found in settlements of the northern Caucasus.[23] Whatever cultural process allowed the people from the south to have such a demographic impact, once the Yamnaya formed, their descendants expanded in all directions.[24]

How the Steppe Came to Central Europe

On the eve of the arrival of steppe ancestry in central Europe around five thousand years ago, the genetic ancestry of the people who lived there was largely derived from the first farmers who had come into Europe from Anatolia beginning after nine thousand years ago, with a minority contribution from the indigenous European hunter-gatherers who mixed with them. In far eastern Europe also around five thousand years ago, the genetic structure of the Yamnaya reflected a different mixture of ancestries: an Iranian-related population along with an eastern European hunter-gatherer population, in approximately equal proportions. Populations that were mixes of

European farmers and steppe groups related to the Yamnaya had not yet formed.

The genetic impact of steppe ancestry on central Europe came in the form of peoples who were part of the ancient culture known to archaeologists as the Corded Ware, so named after its pots decorated by the impressing of twine into soft clay. Beginning around forty-nine hundred years ago, artifacts characteristic of the Corded Ware culture started spreading over a vast region, from Switzerland to European Russia. The ancient DNA data showed that beginning with the Corded Ware culture, individuals with ancestry similar to present-day Europeans first appeared in Europe.[25] Nick Patterson, Iosif Lazaridis, and I developed new statistical methods that allowed us to estimate that in Germany, people buried with Corded Ware pots derive about three-quarters of their ancestry from groups related to the Yamnaya and the rest from people related to the farmers who had been the previous inhabitants of that region. Steppe ancestry has endured, as we also found it in all subsequent archaeological cultures of northern Europe as well as in all present-day northern Europeans.

The genetic data thus settled a long-standing debate in archaeology about linkages between the Corded Ware and the Yamnaya cultures. The two had many striking parallels, such as the construction of large burial mounds, the intensive exploitation of horses and herding, and a strikingly male-centered culture that celebrated violence, as reflected in the great maces (or hammer-axes) buried in some graves. At the same time, there were profound differences between the two cultures, notably the entirely different types of pottery that they made, with important elements of the Corded Ware style adapted from previous central European pottery styles. But the genetics showed that the connection between the Corded Ware culture and the Yamnaya culture reflected major movements of people. The makers of the Corded Ware culture were, at least in a genetic sense, a westward extension of the Yamnaya.

The discovery that the Corded Ware culture reflected a mass migration of people into central Europe from the steppe was not just a sterile academic finding. It had political and historical resonance. At the beginning of the twentieth century, the German archaeologist Gustaf Kossinna was among the first to articulate the idea that cultures of the past that were spread across large geographic regions could be recognized through similarities in style of the artifacts they left behind. He also went further in viewing archaeologically identified cultures as synonymous with peoples, and he originated the idea that the spread of material culture could be used to trace ancient migrations, an approach he called the *siedlungsarchäologische Methode*, or "Settlement Archaeology." Based on the overlap of the geographic distribution of the Corded Ware culture with the places where German is spoken, Kossinna suggested that the cultural roots of the Germans and of Germanic languages today lay in the Corded Ware culture. In his essay "The Borderland of Eastern Germany: Home Territory of the Germans," he argued that because the Corded Ware culture included the territories of Poland, Czechoslovakia, and western Russia of his day, it gave Germans the moral birthright to claim those regions as their own.[26]

Kossinna's ideas were embraced by the Nazis, and although he died in 1931, before they came to power, his scholarship was used as a basis for their propaganda and a justification for their claims to territories to the east.[27] Kossinna's suggestion that migration was the primary explanation for changes in the archaeological record was also attractive to the Nazis because it played into their racist worldview, as it was easy to imagine that migrations had been propelled by innate biological superiority of some peoples over others. Following the Second World War, European archaeologists reacting to the politicization of their field began picking apart the arguments of Kossinna and his colleagues, documenting cases in which changes in material culture were brought about through local invention or imitation and not the spread of people. They urged extreme caution about invoking migration to explain changes in the archaeological record. Today, a common view among archaeologists is that migrations are only one of many explanations for past cultural change. Many archaeologists still argue that when there is evidence for major cultural change at a site, the working assumption should be that the changes reflect communication of ideas or local invention, not necessarily movements of people.[28]

Discussions of the Corded Ware culture and migration in the same breath ring particularly loud alarm bells because of Kossinna's and the Nazis' attempt to use the Corded Ware culture to construct a basis for national German identity.[29] While we were in the final stages of preparing a paper for submission in 2015, one of the German archaeologists who contributed skeletal samples wrote a letter to all coauthors: "We must(!) avoid . . . being compared with the so called 'siedlungsarchäologische Methode' from Gustaf Kossinna!" He and several contributors then resigned as authors before we modified our paper to highlight differences between Kossinna's thesis and our findings, namely that the Corded Ware culture came from the east and that the people associated with it had not been previously established in central Europe.

The correct theory that the Corded Ware culture spread through a migration from the east had already been proposed in the 1920s by Kossinna's contemporary, the archaeologist V. Gordon Childe,[30] although this idea too fell out of favor in the wake of the Second World War and the reaction to the abuse of archaeology by the Nazis, a reaction that took the form of extreme skepticism about any claims of migration.[31] Our finding about the genetic link between the Yamnaya and the Corded Ware culture demonstrates the disruptive power of ancient DNA. It can prove past movements of people, and in this case has documented a magnitude of population replacement that no modern archaeologist, even the most ardent supporter of migrations, had dared to propose. The association between steppe genetic ancestry and people assigned to the Corded Ware archaeological culture through graves and artifacts is not simply a hypothesis. It is now a proven fact.

How was it that the low-population-density shepherds from the steppe were able to displace the densely settled farmers of central and western Europe? The archaeologist Peter Bellwood has argued that once densely settled farming populations were established in Europe, it would have been practically impossible for other groups coming in to make a demographic dent, as their numbers would, he thought, have been dwarfed by the already established population.[32] As an analogy, consider the effect of the British or Mughal occupations of India. Both powers controlled the subcontinent for hundreds of years, but left little trace in the people there today. But ancient DNA shows definitively that major population replacement happened in Europe after around forty-five hundred years ago.

How were people with steppe ancestry able to have such an impact on an already settled region? A possible answer is that the farmers who preceded them may not have occupied every available economic niche in central Europe, giving the steppe peoples an opportunity to expand. Although it is difficult to estimate population sizes from archaeological evidence, the number of people in northern Europe before two thousand years ago has been estimated to be around one hundred times less than today or even smaller, reflecting less efficient farming methods, lack of access to pesticides and fertilizers, the absence of high-yielding plant varieties, and higher infant mortality.[33] When the Corded Ware culture arrived, many tilled fields in central Europe were surrounded by virgin forests. But studies of pollen records in Denmark and elsewhere show that around this time, large parts of northern Europe were transformed from partial forest to grasslands, suggesting that the Corded Ware newcomers may have cut down forests, reengineered parts of the landscape to be more like the steppe, and carved out a niche for themselves that previous peoples of the region had never fully claimed.[34]

There is also a second possible explanation for why the steppe peoples were able to become established in Europe—one that no one would have thought plausible without ancient DNA. Eske Willerslev and Simon Rasmussen, working with the archaeologist Kristian Kristiansen, had the idea of testing 101 ancient DNA samples from Europe and the steppe for evidence of pathogens.[35] In seven samples, they found DNA from *Yersinia pestis*, the bacterium responsible for the Black Death, estimated to have wiped out around one-third of the populations of Europe, India, and China around seven hundred years ago. Traces of plague in a person's teeth are almost a sure sign that he or she died of it. The earliest bacterial genomes that they sequenced lacked a few key genes necessary for the disease to spread via fleas, which is necessary to cause bubonic plague. But the bacterial genomes did carry the genes necessary to cause pneumonic plague, which is spread by sneezing and coughing just like the flu. That a substantial fraction of random graves analyzed carried *Y. pestis* shows that this disease was endemic on the steppe.

Is it possible that the steppe people had picked up the plague and built up an immunity to it, and then transmitted it to the immunologically susceptible central European farmers, causing their numbers to collapse and thereby clearing the way for the Corded Ware culture expansion? This would be a great irony. One of the most important reasons for the collapse of Native American populations after 1492 was infectious diseases spread by Europeans who plausibly had built up some immunity to these diseases after thousands of years of exposure as a result of living in close proximity to their farm animals. But Native Americans, who by and large lacked domesticated animals, likely had much less resistance to them. Was it possible that, in a similar way, northern European farmers after five thousand years ago were decimated by plagues brought from the east, paving the way for the spread of steppe ancestry through Europe?

How Britain Succumbed

After the wave of steppe ancestry crashed over central Europe, it kept rolling. Beginning around forty-seven hundred years ago, a couple of centuries after the Corded Ware culture swept into central Europe, there was an equally dramatic expansion of the Bell Beaker culture, probably from the region of present-day Iberia. The Bell Beaker culture is named for its bell-shaped drinking vessels that rapidly spread over a vast expanse of western Europe alongside other artifacts including decorative buttons and archers' wristguards. It is possible to learn about the movement of people and objects by studying the ratios of isotopes of elements like strontium, lead, and oxygen that are characteristic of materials in different parts of the world. By studying the isotopic composition of teeth, archaeologists have shown that some people of the Bell Beaker culture moved hundreds of kilometers from their places of birth.[36] Bell Beaker culture spread to Britain after forty-five hundred years ago.

A major open question for understanding the spread of the Bell Beaker culture has always been whether it was propelled by the movement of people or the spread of ideas. At the beginning of the twentieth century, the recognition of the massive impact of the Bell Beaker culture led to the romantic notion of a "Beaker Folk," a people who disseminated a new culture and perhaps Celtic languages—a nod to the nationalistic fervor of the time. But, like the claim made for the Corded Ware culture, this position fell out of favor after the Second World War.

In 2017, my laboratory succeeded in assembling whole-genome ancient DNA data from more than two hundred skeletons associated with the Beaker culture from across Europe.[37] Iñigo Olalde, a postdoctoral scientist, analyzed the data to show that individuals in Iberia were genetically indistinguishable from the people who had preceded them and who were not buried in a Bell Beaker culture style. But Bell Beaker–associated individuals in central Europe were extremely different, with most of their ancestry of steppe origin, and little if any ancestry in common with individuals from Iberia associated with the Bell Beaker culture. So, in contrast to what happened with the spread of the Corded Ware culture from the east, the initial spread of the Bell Beaker culture across Europe was mediated by the movement of ideas, not by migration.

Once the Bell Beaker culture reached central Europe through the dispersal of ideas, though, it spread further through migration. Prior to the spread of Beaker culture into Britain, not a single ancient DNA sample from among the many dozen we analyzed had any steppe ancestry. But after forty-five hundred years ago, each one of the many dozens of ancient British samples we analyzed had large amounts of steppe ancestry and no special affinity to Iberians at all. Measured in terms of its proportion of steppe ancestry, DNA extracted from dozens of Bell Beaker skeletons in Britain closely matches that of skeletons from Bell Beaker culture graves across the English Channel. The genetic impact of the spread of peoples from the continent into the British Isles in this period was permanent. British and Irish[38] skeletons from the Bronze Age that followed the Beaker period had at most around 10 percent ancestry from the first farmers of these islands, with the other 90 percent from people like those associated with the Bell Beaker culture in the Netherlands. This was a population replacement at least as dramatic as the one that accompanied the spread of the Corded Ware culture.

It turns out that the discredited idea of the "Beaker Folk" was right for Britain, although wrong as an explanation for the spread of the Bell Beaker culture over the European continent as a whole. So

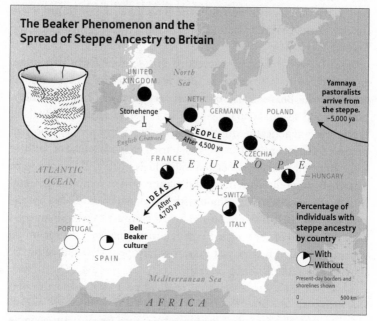

Figure 16. The spread of Beaker pottery between present-day Spain and Portugal and central Europe was due to a movement of ideas, not people, as reflected in their different ancestry patterns. However, the spread of Beaker pottery to the British Isles was accompanied by mass migration. We know this because about 90 percent of the population that built Stonehenge—people with no Yamnaya ancestry—was replaced by people from continental Europe who had such ancestry.

it is that ancient DNA data are beginning to provide us with a more nuanced view of how cultures changed in prehistory. Prompted by the ancient DNA results, several archaeologists have speculated to me that the Bell Beaker culture could be viewed as a kind of ancient religion that converted peoples of different backgrounds to a new way of viewing the world, thus serving as an ideological solvent that facilitated the integration and spread of steppe ancestry and culture into central and western Europe. At a Hungarian Bell Beaker site, we found direct evidence that this culture was open to people of diverse ancestries, with individuals buried in a Bell Beaker cultural context having the full range of steppe ancestry from zero to 75 percent (as high as in people associated with the Corded Ware culture).

What made it possible for people practicing the Beaker culture to spread so dramatically into northwestern Europe and outcompete the established and highly sophisticated populations previously established there? Archaeologists view the Bell Beaker culture as extremely different from the Corded Ware culture, which was in turn extremely different from the Yamnaya culture. Yet all three participated in the massive spread of steppe ancestry from east to west, and perhaps they shared some elements of an ideology despite their very different features.

Speculations about shared features among cultures separated from each other by hundreds of kilometers make scientists and archaeologists uncomfortable. But we should pay attention. Prior to the genetic findings, any claim that a new way of seeing the world could have been shared across cultures as archaeologically different from one another as the Yamnaya, Corded Ware, and Bell Beaker could confidently be dismissed as fanciful. But now we know that these people were linked by major migrations, some of which overwhelmed earlier cultures, providing evidence that these migrations had profound effects. We also need to look again at the spread of language, a direct manifestation of the spread of culture. That almost all Europeans today speak closely related languages is proof that there was strong dissemination of a new culture across Europe at one time. Could the spread of shared languages across Europe have been propelled by the spread of people documented by ancient DNA?

The Origin of Indo-European Languages

A great mystery of prehistory is the origin of Indo-European languages, the closely related group of tongues that today are spoken across almost all of Europe, Armenia, Iran, and northern India, with a great gap in the Near East where these languages only existed at the periphery for the last five thousand years—a fact known to us because writing was invented there.

One of the first people to note the similarity among Indo-European languages was William Jones, a judge serving in Kolkata in British India, who knew Greek and Latin from his schooldays, and had learned Sanskrit, the language of the ancient Indian religious texts. In 1786, he observed: "The Sanskrit language, whatever may be its antiquity, is of a wonderful structure; more perfect than the Greek, more copious than the Latin, and more exquisitely refined than either, yet bearing to both of them a stronger affinity, both in the roots of verbs and in the forms of grammar, than could possibly have been produced by accident; so strong indeed that no philologer could examine them all three, without believing them to have sprung from some common source, which, perhaps, no longer exists."[39] For more than two hundred years, scholars have puzzled over how such a similarity of languages developed over so vast a region.

In 1987, Colin Renfrew proposed a unified theory for how Indo-European languages attained their current distribution. In his book *Archaeology and Language: The Puzzle of Indo-European Origins*, he suggested that the homogeneity of language across such a vast stretch of Eurasia today could be explained by one and the same event: the spread from Anatolia after nine thousand years ago of peoples bringing agriculture.[40] His argument was rooted in the idea that farming would have given Anatolians an economic advantage that would have allowed new populations to spread massively into Europe. Anthropological studies have consistently shown that major migrations of people are necessary to achieve language change in small-scale societies, so a phenomenon as profound as the spread of Indo-European languages was likely to have been propelled by mass migration.[41] Since there was no good archaeological evidence for a later major migration into Europe, and since once densely settled farming populations were established it was difficult to imagine how other groups could gain a foothold, Renfrew and scholars who followed him concluded that the spread of farming was probably what brought Indo-European languages to Europe.[42]

Renfrew's logic was compelling given the data he had available at the time, but the argument that the spread of farming from Anatolia drove the spread of Indo-European languages into Europe has been undermined by the findings from studies of ancient DNA, which showed that a mass movement of people into central Europe occurred after five thousand years ago in association with the Corded Ware culture. By arguing from first principles—that after the spread of farming into Europe it would not have been demographically plausible for there to have been another migration substantial enough to induce a language shift—Renfrew constructed a compelling case for the Anatolian hypothesis, which won many adherents. But theory is always trumped by data, and the data show that the Yamnaya also made a major demographic impact—in fact, it is clear that the single most important source of ancestry across northern Europe today is the Yamnaya or groups closely related to them. This suggests that the Yamnaya expansion likely spread a major new group of languages throughout Europe. The ubiquity of Indo-European languages in Europe over the last few thousand years, and the fact that the Yamnaya-related migration was more recent than the farming one, makes it likely that at least some Indo-European languages in Europe, and perhaps all of them, were spread by the Yamnaya.[43]

The main counterargument to the Anatolian hypothesis is the steppe hypothesis—the idea that Indo-European languages spread from the steppe north of the Black and Caspian seas. The best single argument for the steppe hypothesis prior to the availability of genetic data may be the one constructed by David Anthony, who has shown that the shared vocabulary of the great majority of present-day Indo-European languages is unlikely to be consistent with their having originated much earlier than about six thousand years ago. His key observation is that all extant branches of the Indo-European language family except for the most anciently diverging Anatolian ones that are now extinct (such as ancient Hittite) have an elaborate shared vocabulary for wagons, including words for axle, harness pole, and wheels. Anthony interpreted this sharing as evidence that all Indo-European languages spoken today, from India in the east to the Atlantic fringe in the west, descend from a language spoken by an ancient population that used wagons. This population could not have lived much earlier than about six thousand years ago, since we know from archaeological evidence that it was around then that wheels and wagons spread.[44] This date rules out the Anatolian farming expansion into Europe between nine thousand and eight thousand years ago. The obvious candidate for dispersing most of today's Indo-European languages is thus the Yamnaya, who depended on the technology of wagons and wheels that became widespread around five thousand years ago.

That there could have been a massive enough migration by steppe pastoralists to displace settled agricultural populations, and thereby distribute a new language, seems on the face of it even more implausible for India than for Europe. India is protected from the steppe by the high mountains of Afghanistan, whereas there is no similar barrier protecting Europe. Yet the steppe pastoralists broke through to India too. As is related in the next chapter, almost everyone in India is a mixture of two highly divergent ancestral populations, one of which derived about half its ancestry directly from the Yamnaya.

While the genetic findings point to a central role for the Yamnaya in spreading Indo-European languages, tipping the scales definitively in favor of some variant of the steppe hypothesis, those findings do not yet resolve the question of the homeland of the original Indo-European languages, the place where these languages were spoken before the Yamnaya so dramatically expanded. Anatolian languages known from four-thousand-year-old tablets recovered from the Hittite Empire and neighboring ancient cultures did not share the full wagon and wheel vocabulary present in all Indo-European languages spoken today. Ancient DNA available from this time in Anatolia shows no evidence of steppe ancestry similar to that in the Yamnaya (although the evidence here is circumstantial as no ancient DNA from the Hittites themselves has yet been published). This suggests to me that the most likely location of the population that first spoke an Indo-European language was south of the Caucasus Mountains, perhaps in present-day Iran or Armenia, because ancient DNA from people who lived there matches what we would expect for a source population both for the Yamnaya and for ancient Anatolians. If this scenario is right, the population sent one branch up into the steppe—mixing with steppe hunter-gatherers in a one-to-one ratio to become the Yamnaya as described earlier—and another to Anatolia to found the ancestors of people there who spoke languages such as Hittite.

To an outsider, it might seem surprising that DNA can have a definitive impact on a debate about language. DNA cannot of course reveal what languages people spoke. But what genetics can do is to establish that migrations occurred. If people moved, it means that cultural contact occurred too—in other words, genetic tracing of migrations makes it possible also to trace potential spreads of culture and language. By tracing possible migration paths and ruling out others, ancient DNA has ended a decades-old stalemate in the controversy regarding the origins of Indo-European languages. The Anatolian hypothesis has lost its best evidence, and the most common version of the steppe hypothesis—which suggests that the ultimate origin of all Indo-European languages including ancient Anatolian languages was in the steppe—has to be modified too. DNA has emerged as central to the new synthesis of genetics, archaeology, and linguistics that is now replacing outdated theories.

A great lesson of the ancient DNA revolution is that its findings almost always provide accounts of human migrations that are very different from preexisting models, showing how little we really knew about human migrations and population formation prior to the invention of this new technology. The vision of Indo-Europeans or "Aryans" as a "pure" group has sparked nationalist sentiments in Europe since the nineteenth century.[45] There were debates about whether the Celts or the Teutons or other groups were the real "Aryans," and Nazi racism was fueled by this discussion. The genetic data have provided what might seem like uncomfortable support for some of these ideas—suggesting that a single, genetically coherent group was responsible for spreading many Indo-European languages. But the data also reveal that these early discussions were misguided in supposing purity of ancestry. Whether the original Indo-European speakers lived in the Near East or in eastern Europe, the Yamnaya, who were the main group responsible for spreading Indo-European languages across a vast span of the globe, were formed by mixture. The people who practiced the Corded Ware culture were a further mixture, and northwestern Europeans associated with the Bell Beaker culture were yet a further mixture. Ancient DNA has established major migration and mixture between highly divergent populations as a key force shaping human prehistory, and ideologies that seek a return to a mythical purity are flying in the face of hard science.

South Asian Population History

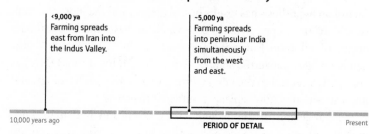

‹9,000 ya
Farming spreads
east from Iran into
the Indus Valley.

~5,000 ya
Farming spreads
into peninsular India
simultaneously
from the west
and east.

10,000 years ago   **PERIOD OF DETAIL**   Present

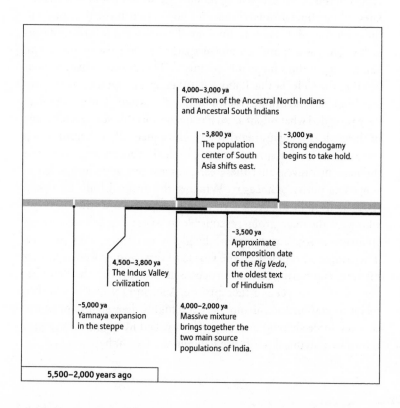

4,000–3,000 ya
Formation of the Ancestral North Indians
and Ancestral South Indians

~3,800 ya
The population
center of South
Asia shifts east.

~3,000 ya
Strong endogamy
begins to take hold.

~3,500 ya
Approximate
composition date
of the *Rig Veda*,
the oldest text
of Hinduism

4,500–3,800 ya
The Indus Valley
civilization

~5,000 ya
Yamnaya expansion
in the steppe

4,000–2,000 ya
Massive mixture
brings together the
two main source
populations of India.

5,500–2,000 years ago

The Collision That Formed India

The Fall of the Indus Civilization

In the oldest text of Hinduism, the *Rig Veda*, the warrior god Indra rides against his impure enemies, or *dasa*, in a horse-drawn chariot, destroys their fortresses, or *pur*, and secures land and water for his people, the *arya*, or Aryans.[1]

Composed between four thousand and three thousand years ago in Old Sanskrit, the *Rig Veda* was passed down orally for some two thousand years before being written down, much like the *Iliad* and *Odyssey* in Greece, which were composed several hundred years later in another early Indo-European language.[2] The *Rig Veda* is an extraordinary window into the past, as it provides a glimpse of what Indo-European culture might have been like in a period far closer in time to when these languages radiated from a common source. But what did the stories of the *Rig Veda* have to do with real events? Who were the *dasa*, who were the *arya*, and where were the fortresses located? Did anything like this really happen?

There was tremendous excitement about the possibility of using archaeology to gain insight into these questions in the 1920s and 1930s. In those years, excavations uncovered the remains of an ancient civilization, walled cities at Harappa, Mohenjo-daro, and elsewhere in the Punjab and Sind that dated from forty-five hundred to thirty-eight hundred years ago. These cities and smaller towns and villages dotted the valley of the river Indus in present-day Pakistan and parts of India, and some of them sheltered tens of thousands of people.[3] Were they perhaps the fortresses, or *pur*, of the *Rig Veda*?

Indus Valley Civilization cities were surrounded by perimeter walls and laid out on grids. They had ample storage for grain supplied by farming of land in the surrounding river plains. The cities sheltered craftspeople skilled in working clay, gold, copper, shell, and wood. The people of the Indus Valley Civilization engaged in prolific trade and commerce, as reflected in the stone weights and measures they left behind, and their trading partners, who lived as far away as Afghanistan, Arabia, Mesopotamia, and even Africa.[4] They made decorative seals with images of humans or animals. There were often signs or symbols on the seals whose meaning remains largely undeciphered.[5]

Since the original excavations, many things about the Indus Valley Civilization have remained enigmatic, not only its script. The greatest mystery is its decline. Around thirty-eight hundred years ago, the settlements of the Indus dwindled, with population centers shifting east toward the Ganges plain.[6] Around this time, the *Rig Veda* was composed in Old Sanskrit, a language that is ancestral to the great majority of languages spoken in northern India today and that had diverged in the millennium before the *Rig Veda* was composed from the languages spoken in Iran. Indo-Iranian languages are in turn cousins of almost all of the languages spoken in Europe and with them make up the great Indo-European language family. The religion of the *Rig Veda*, with its pantheon of deities governing nature and regulating society, had unmistakable similarities to the mythology of other parts of Indo-European Eurasia, including Iran, Greece, and Scandinavia, providing further evidence of cultural links across vast expanses of Eurasia.[7]

Some have speculated that the collapse of the Indus Valley Civilization was caused by the arrival in the region of migrants from the north and west speaking Indo-European languages, the so-called Indo-Aryans. In the *Rig Veda*, the invaders had horses and chariots. We know from archaeology that the Indus Valley Civilization was a pre-horse society. There is no clear evidence of horses at their sites, nor are there remains of spoke-wheeled vehicles, although there are clay figurines of wheeled carts pulled by cattle.[8] Horses and spoke-wheeled chariots were the weapons of mass destruction of Bronze Age Eurasia. Did the Indo-Aryans use their military technology to put an end to the old Indus Valley Civilization?

Since the original excavations at Harappa, the "Aryan invasion theory" has been seized on by nationalists in both Europe and India, which makes the idea difficult to discuss in an objective way. European racists, including the Nazis, were drawn to the idea of an invasion of India in which the dark-skinned inhabitants were subdued by light-skinned warriors related to northern Europeans, who imposed on them a hierarchical caste system that forbade intermarriage across groups. To the Nazis and others, the distribution of the Indo-European language family, linking Europe to India and having little impact on the Near East with its Jews, spoke of an ancient conquest moving out of an ancestral homeland, displacing and subjugating the peoples of the conquered territories, an event that they wished to emulate.[9] Some placed the ancestral homeland of the Indo-Aryans in northeast Europe, including Germany. They also adopted features of Vedic mythology as their own, calling themselves Aryans after the term in the *Rig Veda*, and appropriating the swastika, a traditional Hindu symbol of good fortune.[10]

The Nazis' interest in migrations and the spread of Indo-European languages has made it difficult for serious scholars in Europe to discuss the possibility of migrations spreading Indo-European languages.[11] In India, the possibility that the Indus Valley Civilization fell at the hands of migrating Indo-European speakers coming from the north is also fraught, as it suggests that important elements of South Asian culture might have been influenced from the outside.

The idea of a mass migration from the north has fallen out of favor among scholars not only because it has become so politicized, but also because archaeologists have realized that major cultural shifts in the archaeological record do not always imply major migrations. And, in fact, there is scant archaeological evidence for such a population movement. There are no obvious layers of ash and destruction around thirty-eight hundred years ago suggesting the burning and sacking of the Indus towns. If anything, there is evidence that the Indus Valley Civilization's decline played out over a long period, with emigration away from the towns and environmental degradation taking place over decades. But the lack of archaeological evidence does not mean that there were no major incursions from the outside. Between sixteen hundred and fifteen hundred years ago, the western Roman Empire collapsed under the pressure of the German expansions, with great political and economic blows dealt to the western Roman Empire when the Visigoths and the Vandals each sacked Rome and took political control of Roman provinces. However, there so far seems to be little archaeological evidence for destruction of Roman cities in this time, and if not for the detailed historical accounts, we might not know these pivotal events occurred.[12] It is possible that in the apparent depopulation of the Indus Valley, too, we might be limited by the difficulty archaeologists have in detecting sudden change. The patterns evident from archaeology may be obscuring more sudden triggering events.

What can genetics add? It cannot tell us what happened at the end of the Indus Valley Civilization, but it can tell us if there was a collision of peoples with very different ancestries. Although mixture is not by itself proof of migration, the genetic evidence of mixture proves that dramatic demographic change and thus opportunity for cultural exchange occurred close to the time of the fall of Harappa.

A Land of Collisions

The great Himalayas were formed around ten million years ago by the collision of the Indian continental plate, moving northward through the Indian Ocean, with Eurasia. India today is also the product of collisions of cultures and people.

Consider farming. The Indian subcontinent is one of the breadbaskets of the world—today it feeds a quarter of the world's population—and it has been one of the great population centers ever since modern humans expanded across Eurasia after fifty thousand years ago. Yet farming was not invented in India. Indian farming today is born of the collision of the two great agricultural systems of Eurasia. The Near Eastern winter rainfall crops, wheat and barley, reached the Indus Valley sometime after nine thousand years ago according to archaeological evidence—as attested, for example, in ancient Mehrgarh on the western edge of the Indus Valley in present-day Pakistan.[13] Around five thousand years ago, local farmers succeeded in breeding these crops to adapt to monsoon summer rainfall patterns, and the crops spread into peninsular India.[14] The Chinese monsoon summer rainfall crops of rice and millet also reached peninsular India around five thousand years ago. India may have been the first place where the Near Eastern and the Chinese crop systems collided.

Language is another blend. The Indo-European languages of the north of India are related to the languages of Iran and Europe. The Dravidian languages, spoken mostly by southern Indians, are not closely related to languages outside South Asia. There are also Sino-Tibetan languages spoken by groups living in the mountains fringing the north of India, and small pockets of tribal groups in the east and center that speak Austroasiatic languages related to Cambodian and Vietnamese, and that are thought to descend from the languages spoken by the peoples who first brought rice farming to South Asia and parts of Southeast Asia. Words borrowed from ancient Dravidian and Austroasiatic languages, which linguists can detect as they are not typical of Indo-European languages, are present in the *Rig Veda*, implying that these languages have been in contact in India for at least three or four thousand years.[15]

The people of India are also diverse in appearance, providing visual testimony to mixture. A stroll down a street in any Indian city makes it clear how diverse Indians are. Skin shades range from dark to pale. Some people have facial features like Europeans, others closer to Chinese. It is tempting to think that these differences reflect a collision of peoples who mixed at some point in the past, with different proportions of mixture in different groups living today. But it is also possible to overinterpret physical appearances, as it is known that appearances can also reflect environment and diet.

The first genetic work in India gave seemingly contradictory results. Researchers studying mitochondrial DNA, always passed down from mothers, found that the vast majority of mitochondrial DNA in Indians was unique to the subcontinent, and they estimated that the Indian mitochondrial DNA types only shared common ancestry with ones predominant outside South Asia many tens of thousands of years ago.[16] This suggested that on the maternal line, Indian ancestors had been largely isolated within the subcontinent for a long time, without mixing with neighboring populations to the west, east, or north. In contrast, a good fraction of Y chromosomes in India, passed from father to son, showed closer relatedness to West Eurasians—Europeans, central Asians, and Near Easterners—suggesting mixture.[17]

Some historians of India have thrown up their hands and discounted genetic information due to these apparently conflicting findings. The situation has not been helped by the fact that geneticists do not have formal training in archaeology, anthropology, and linguistics—the fields that have dominated the study of human prehistory—and are prone to make elementary mistakes or to be tripped up by known fallacies when summarizing findings from those fields. But it is foolhardy to ignore genetics. We geneticists may be the barbarians coming late to the study of the human past, but it is always a bad idea to ignore barbarians. We have access to a type of data that no one has had before, and we are wielding these data to address previously unapproachable questions about who ancient peoples were.

The Isolated People of Little Andaman Island

My research into the prehistory of India began in 2007 with a book and a letter.

The book was *The History and Geography of Human Genes*, Luca Cavalli-Sforza's magnum opus, in which he mentions the "Negrito" people of the Andaman Islands in the Bay of Bengal, hundreds of kilometers from the mainland. The Andaman Islands have remained isolated by deep sea barriers for most of the history of modern human dispersal through Eurasia, although the largest, Great Andaman, has been massively disrupted by mainland influence over the last few hundred years (the British used it as a colonial prison). North Sentinel Island is populated by one of the last largely uncontacted

Stone Age peoples of the world—a group of several hundred people who are now protected from outside interference by the Indian
government, and who are so not-of-our-world that they shot arrows
at Indian helicopters sent to offer help after the Indian Ocean tsunami of 2004. The Andamanese speak languages that are so different
from any others in Eurasia that they have no traceable connections.
They also look very different from other humans living nearby, with
slighter frames and tightly coiled hair. In one section of his book,
Cavalli-Sforza speculated that the Andamanese might represent
isolated descendants of the earliest expansions of modern humans
out of Africa, perhaps having moved there before the migration that
occurred after around fifty thousand years ago and that gave rise to
most of the ancestry of non-Africans today.

On reading this, my colleagues and I wrote a letter to Lalji Singh
and Kumarasamy Thangaraj of the Centre for Cellular and Molecular Biology (CCMB) in Hyderabad, India. A few years earlier,
Singh and Thangaraj had published a paper on mitochondrial and
Y-chromosome DNA from people of the Andaman Islands.[18] Their
study showed that the people of Little Andaman Island had been
separated for tens of thousands of years from peoples of the Eurasian mainland. I asked them whether it would be possible to analyze
whole genomes of the Andamanese, to gain a fuller picture.

Singh and Thangaraj were excited to collaborate and quickly convinced me that there was a broader picture to paint involving mainland Indians as well. They offered us access to a vast collection of
DNA. In the freezers at CCMB, they had assembled samples that
represented the extraordinary human diversity of India—the last time
I checked, the collection included more than three hundred groups
and more than eighteen thousand individual DNA samples. These
had been assembled by students from all over India who had visited
villages and collected blood samples from people whose grandparents
were from the same location and group. From the CCMB collection, we selected twenty-five groups that were as diverse as possible
geographically, culturally, and linguistically. The groups were of
traditionally high as well as low social status in the Indian caste system, and also included a number of tribes entirely outside the caste
system.

A few months later, Thangaraj came to our laboratory in Boston,

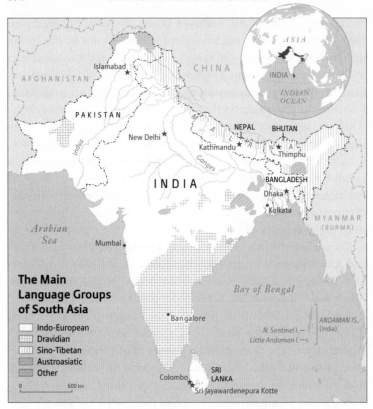

Figure 17a. People in the north primarily speak Indo-European languages and have relatively high proportions of West Eurasian–related ancestry. People in the south primarily speak Dravidian languages and have relatively low proportions of West Eurasian ancestry. Many groups in the north and east speak Sino-Tibetan languages. Isolated tribal groups in the center and east speak Austroasiatic languages.

bringing with him this unique and precious set of DNA samples. We analyzed them using a single nucleotide polymorphism (SNP) microarray, a technology that had just recently become available in the United States but was not yet available in India. For this reason, Thangaraj had been granted permission by the Indian government to take the DNA outside India. (There are Indian regulations limiting export of biological material if the research can be achieved within the country.)

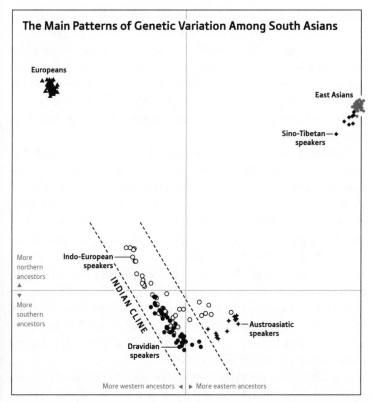

Figure 17b. Analysis of the primary patterns of genetic variation in South Asia shows that the majority of Indian groups form a gradient of ancestry, with Indo-European speakers from the north clustering at one extreme, and Dravidian speakers from the south at the other.

A SNP microarray contains hundreds of thousands of microscopic pixels, each of which is covered by artificially synthesized stretches of DNA from the places in the genome that scientists have chosen to analyze. When a DNA sample is washed over the microarray, the fragments that overlap the artificial DNA sequences bind tightly, and the fragments that do not are washed away. Based on the relative intensity of binding to these bait sequences, a camera that detects fluorescent light can determine which possible genetic types a person carries in his or her genome. The SNP microarray that we analyzed was able to study many hundreds of thousands of positions in the genome that harbor a mutation carried by some people but not others. By studying these positions, it is possible to determine which people are most closely related to which others. The technique is much less expensive than sequencing a whole human genome since it zeroes in on points of interest—those that tend to differ among people and thus provide the greatest density of information about population history.

To obtain an initial picture of how the samples were related to each other, we used the mathematical technique of principal component analysis, which is also described in the previous chapter on West Eurasian population history, and which finds combinations of single-letter changes in DNA that are most informative about the differences among people. Using this method to display Indian genetic data on a two-dimensional graph, we found that the samples spread out along a line. At the far extreme of the line were West Eurasian individuals—Europeans, central Asians, and Near Easterners—which we had included in the analysis for the sake of comparison. We called the non–West Eurasian part of the line the "Indian Cline": a gradient of variation among Indian groups that pointed on the plot like an arrow directly at West Eurasians.[19]

A gradient in a principal component analysis plot can be caused by several quite different histories, but such a striking pattern led us to guess that many Indian groups today might be mixtures, in different proportions, of a West Eurasian–related ancestral population and another very different population. Seeing that the southernmost groups in India—which also spoke Dravidian languages—tended to be farthest away from West Eurasians in the plot, we explored a model in which Indians today are formed from a mixture of two ancestral populations, and we evaluated the consistency of this model with the data.

To test whether mixture occurred, we had to develop new methods. The methods that we applied in 2010 to show that mixture had occurred between Neanderthals and modern humans[20] were in fact primarily developed to study Indian population history.

We first tested the hypothesis that Europeans and Indians descend from a common ancestral population that split at an earlier time from the ancestors of East Asians such as Han Chinese. We identified DNA letters where European and Indian genomes differed, and then measured how often Chinese samples had the genetic types seen in Europeans or Indians. We found that Chinese clearly share more DNA letters with Indians than they do with Europeans. That ruled out the possibility that Europeans and Indians descended from a common homogeneous ancestral population since their separation from the ancestors of Chinese.

We then tested the alternative hypothesis that Chinese and Indians descend from a common ancestral population since their separation from the ancestors of Europeans. However, this scenario did not hold up either: European groups are more closely related to all Indians than to all Chinese.

We found that the frequencies of the genetic mutations seen in all Indians are, on average, intermediate between those in Europeans and East Asians. The only way that this pattern could arise was through mixture of ancient populations—one related to Europeans, central Asians, and Near Easterners, and another related distantly to East Asians.

We initially called the first population "West Eurasians," as a way of referring to the large set of populations in Europe, the Near East, and central Asia, among which there are only modest differences in the frequencies of genetic mutations from one group to another. These differences are typically about ten times smaller than the differences between Europeans and the people of East Asia. It was striking to find that one of the two populations contributing to the ancestry of Indians today grouped with West Eurasians. This looked to us like the easternmost edge of the ancient distribution of West Eurasian ancestry, where it had mixed with other very different people. We could see that the other population was more closely related to present-day East Asians such as Chinese, but was also clearly tens of thousands of years separated from them. So it represented an early-diverging lineage that contributed to people living today in South Asia but not much to people living anywhere else.

Having identified the mixture, we searched for present-day Indian populations that might have escaped it. All the populations on the mainland had some West Eurasian–related ancestry. However, the people of Little Andaman Island had none. The Andamanese were consistent with being isolated descendants of an ancient East Asian–related population that contributed to South Asians. The indigenous people of Little Andaman Island, despite a census size of fewer than one hundred, turned out to be key to understanding the population history of India.

The Mixing of East and West

The tensest twenty-four hours of my scientific career came in October 2008, when my collaborator Nick Patterson and I traveled to Hyderabad to discuss these initial results with Singh and Thangaraj.

Our meeting on October 28 was challenging. Singh and Thangaraj seemed to be threatening to nix the whole project. Prior to the meeting, we had shown them a summary of our findings, which were that Indians today descend from a mixture of two highly divergent ancestral populations, one being "West Eurasians." Singh and Thangaraj objected to this formulation because, they argued, it implied that West Eurasian people migrated en masse into India. They correctly pointed out that our data provided no direct evidence for this conclusion. They even reasoned that there could have been a migration in the other direction, of Indians to the Near East and Europe. Based on their own mitochondrial DNA studies, it was clear to them that the great majority of mitochondrial DNA lineages present in India today had resided in the subcontinent for many tens of thousands of years.[21] They did not want to be part of a study that suggested a major West Eurasian incursion into India without being absolutely certain as to how the whole-genome data could be reconciled with their mitochondrial DNA findings. They also implied that the suggestion of a migration from West Eurasia would be politically explosive. They did not explicitly say this, but it had obvious overtones of the idea that migration from outside India had a transformative effect on the subcontinent.

Singh and Thangaraj suggested the term "genetic sharing" to describe the relationship between West Eurasians and Indians, a formulation that could imply common descent from an ancestral population. However, we knew from our genetic studies that a real and profound mixture between two different populations had occurred and made a contribution to the ancestry of almost every Indian living today, while their suggestion left open the possibility that no mixture had happened. We came to a standstill. At the time I felt that we were being prevented by political considerations from revealing what we had found.

That evening, as the fireworks of Diwali, one of the most important holidays of the Hindu year, crackled, and as young boys threw sparklers beneath the wheels of moving trucks outside our compound, Patterson and I holed up in his guest room at Singh and Thangaraj's scientific institute and tried to understand what was going on. The cultural resonances of our findings gradually became clear to us. So we groped toward a formulation that would be scientifically accurate as well as sensitive to these issues.

The next day, the full group reconvened in Singh's office. We sat together and came up with new names for ancient Indian groups. We wrote that the people of India today are the outcome of mixtures between two highly differentiated populations, "Ancestral North Indians" (ANI) and "Ancestral South Indians" (ASI), who before their mixture were as different from each other as Europeans and East Asians are today. The ANI are related to Europeans, central Asians, Near Easterners, and people of the Caucasus, but we made no claim about the location of their homeland or any migrations. The ASI descend from a population not related to any present-day populations outside India. We showed that the ANI and ASI had mixed dramatically in India. The result is that everyone in mainland India today is a mix, albeit in different proportions, of ancestry related to West Eurasians, and ancestry more closely related to diverse East Asian and South Asian populations. No group in India can claim genetic purity.

Ancestry, Power, and Sexual Dominance

Having come to this conclusion, we were able to estimate the fraction of West Eurasian–related ancestry in each Indian group.

To make these estimates, we measured the degree of the match of a West Eurasian genome to an Indian genome on the one hand and to a Little Andaman Islander genome on the other. The Little Andamanese were crucial here because they are related (albeit distantly) to the ASI but do not have the West Eurasian–related ancestry present in all mainland Indians, so we could use them as a reference point for our analysis. We then repeated the analysis, now replacing the Indian genome with the genome of a person from the Caucasus to measure the match rate we should expect if a genome was entirely of West Eurasian–related ancestry. By comparing the two numbers, we could ask: "How far is each Indian population from what we would expect for a population of entirely West Eurasian ancestry?" By answering this question we could estimate the proportion of West Eurasian–related ancestry in each Indian population.

In this initial study and in subsequent studies with larger numbers of Indian groups, we found that West Eurasian–related mixture in India ranges from as low as 20 percent to as high as 80 percent.[22] This continuum of West Eurasian–related ancestry in India is the reason for the Indian Cline—the gradient we had seen on our principal components plots. No group is unaffected by mixing, neither the highest nor the lowest caste, including the non-Hindu tribal populations living outside the caste system.

The mixture proportions provided clues about past events. For one thing, the genetic data hinted at the languages spoken by the ancient ANI and ASI. Groups in India that speak Indo-European languages typically have more ANI ancestry than those speaking Dravidian languages, who have more ASI ancestry. This suggested to us that the ANI probably spread Indo-European languages, while the ASI spread Dravidian languages.

The genetic data also hinted at the social status of the ancient ANI (higher social status on average) and ASI (lower social status on average). Groups of traditionally higher social status in the Indian caste system typically have a higher proportion of ANI ancestry than those of traditionally lower social status, even within the same state of India where everyone speaks the same language.[23] For example, Brahmins, the priestly caste, tend to have more ANI ancestry than the groups they live among, even those speaking the same language. Although there are groups in India that are exceptions to these patterns, including well-documented cases where whole groups have shifted social status,[24] the findings are statistically clear, and suggest that the ANI-ASI mixture in ancient India occurred in the context of social stratification.

The genetic data from Indians today also reveal something about the history of differences in social power between men and women. Around 20 to 40 percent of Indian men and around 30 to 50 percent of eastern European men have a Y-chromosome type that, based on the density of mutations separating people who carry it, descends in the last sixty-eight hundred to forty-eight hundred years from the same male ancestor.[25] In contrast, the mitochondrial DNA, passed down along the female line, is almost entirely restricted to India, suggesting that it may have nearly all come from the ASI, even in the north. The only possible explanation for this is major migration between West Eurasia and India in the Bronze Age or afterward. Males with this Y chromosome type were extraordinarily successful at leaving offspring while female immigrants made far less of a genetic contribution.

The discrepancy between the Y-chromosome and mitochondrial DNA patterns initially confused historians.[26] But a possible explanation is that most of the ANI genetic input into India came from males. This pattern of sex-asymmetric population mixture is disturbingly familiar. Consider African Americans. The approximately 20 percent of ancestry that comes from Europeans derives in an almost four-to-one ratio from the male side.[27] Consider Latinos from Colombia. The approximately 80 percent of ancestry that comes from Europeans is derived in an even more unbalanced way from males (a fifty-to-one ratio).[28] I explore in part III what this means for the relationships among populations, and between males and females, but the common thread is that males from populations with more power tend to pair with females from populations with less. It is amazing that genetic data can reveal such profound information about the social nature of past events.

Population Mixture at the Twilight of Harappa

To understand what our findings about population mixture meant in the context of Indian history, we needed to know not just that population mixture had occurred, but also when.

One possibility we considered is that the mixtures we had detected were due to great human migrations at the end of the last ice age, after around fourteen thousand years ago, as improving climates changed deserts into habitable land and contributed to other environmental change that drove people hither and yon across the landscape of Eurasia.

A second possibility is that the mixtures reflected movements of farmers of Near Eastern origin into South Asia, a migration that could be a possible explanation for the spread of Near Eastern farming into the Indus Valley after nine thousand years ago.

A third possibility is that the mixtures occurred in the last four thousand years associated with the dispersal of Indo-European languages that are spoken today in India as well as in Europe. This possibility hints at events described in the *Rig Veda*. However, even if mixture occurred after four thousand years ago, it is entirely possible that it took place between already-resident populations, one of which had migrated to the area from West Eurasia some centuries or even millennia earlier but had not yet interbred with the ASI.

All three of the possibilities involve migration at some point from West Eurasia into India. Although Singh and Thangaraj entertained the possibility of a migration out of India and into points as far west as Europe to explain the relatedness between the ANI and West Eurasian populations, I have always thought, based on the absence of any trace of ASI ancestry in the great majority of West Eurasians today and the extreme geographic position of India within the present-day distribution of peoples bearing West Eurasian–related ancestry, that the shared ancestry likely reflected ancient migrations into South Asia from the north or west. By dating the mixture, we could obtain more concrete information.

The challenge of getting a date prompted us to develop a series of new methods. Our approach was to take advantage of the fact that in the first generation after the ANI and ASI mixed, their offspring would have had chromosomes of entirely ANI or ASI ancestry. In each subsequent generation, as individuals combined their mother's and father's chromosomes to produce the chromosome they passed on to their offspring, the stretches of ANI and ASI ancestry would have broken up, with one or two breakpoints per generation per chromosome. By measuring the typical size of stretches of ANI or ASI ancestry in Indians today, and determining how many generations would be needed to chop them down to their current size, Priya Moorjani, a graduate student in my laboratory, succeeded in estimating a date.[29]

We found that all Indian groups we analyzed had ANI-ASI mixture dates between four thousand and two thousand years ago, with Indo-European-speaking groups having more recent mixture dates on average than Dravidian-speaking groups. The older mixture dates in Dravidian speakers surprised us. We had expected that the oldest mixtures would be found in Indo-European-speaking groups of the north, as it is presumably there that the mixture first occurred. We then realized that an older date in Dravidians actually makes sense, as the present-day locations of people do not necessarily reflect their past locations. Suppose that the first round of mixture in India happened in the north close to four thousand years ago, and was followed by subsequent waves of mixture in northern India as previously established populations and people with much more West Eurasian ancestry came into contact repeatedly along a boundary zone. The people who were the products of the first mixtures in northern India could plausibly, over thousands of years, have mixed with or migrated to southern India, and thus the dates in southern Indians today would be those of the first round of mixture. Later waves of mixture of West Eurasian–related people into northern Indian groups would then cause the average date of mixture estimated in northern Indians today to be more recent than in southern Indians.

A hard look at the genetic data confirms the theory of multiple waves of ANI-related mixture into the north. Interspersed among the short stretches of ANI-derived DNA we find in northern Indians, we also find quite long stretches of ANI-derived DNA, which must reflect recent mixtures with people of little or no ASI ancestry.[30]

Remarkably, the patterns we observed were consistent with the hypothesis that all of the mixture of ANI and ASI ancestry that occurred in the history of some present-day Indian groups happened within the last four thousand years. This meant that the population structure of India before around four thousand years ago was profoundly different from what it is today. Before then, there were unmixed populations, but afterward, there was convulsive mixture in India, which affected nearly every group.

So between four thousand and three thousand years ago—just as the Indus Civilization collapsed and the *Rig Veda* was composed—there was a profound mixture of populations that had previously been segregated. Today in India, people speaking different languages and coming from different social statuses have different proportions of ANI ancestry. Today, ANI ancestry in India derives more from males than from females. This pattern is exactly what one would expect from an Indo-European-speaking people taking the reins of political and social power after four thousand years ago and mixing with the local peoples in a stratified society, with males from the groups in power having more success in finding mates than those from the disenfranchised groups.

The Antiquity of Caste

How is it that the genetic marks of these ancient events have not been blurred beyond recognition after thousands of years of history?

One of the most distinctive features of traditional Indian society is caste—the system of social stratification that determines whom one can marry and what privileges and roles one has in society. The repressive nature of caste has spawned in reaction major religions—Jainism, Buddhism, and Sikhism—each of which offered refuge from the caste system. The success of Islam in India was also fueled by the escape it provided for low-social-status groups that converted en masse to the new religion of the Mughal rulers. Discrimination on the basis of caste was outlawed with the birth of democratic India, but it still shapes whom people choose to socialize with and marry today.

A sociological definition of a caste is a group that interacts economically with people outside it (through specialized economic roles), but segregates itself socially through endogamy (which prevents people from marrying outsiders). Jews in northeastern Europe, from whom I descend, were, prior to the "Jewish emancipation" beginning in the late eighteenth century, a caste in lands where not all groups were castes. Jews served an economic function as moneylenders, liquor vendors, merchants, and craftspeople for the population within which they lived. Religious Jews then as now segregated themselves socially through dietary rules (kosher laws), distinctive dress, body modification (circumcision of males), and strictures against marrying outsiders.

Caste in India is organized at two levels, *varna* and *jati*.[31] The *varna* system involves stratification of all of society into at least four ranks: at the top the priestly group (Brahmins) and the warrior group (Kshatriyas); in the middle the merchants, farmers, and artisans (Vaishya); and finally the lower castes (Shudras), who are laborers. There are also the Chandalas or Dalits—"Scheduled Castes"— people who are considered so low that they are "untouchable" and excluded from normal society. Finally, there are the "Scheduled Tribes," the official Indian government name for people outside Hinduism who are neither Muslim nor Christian. The caste system is a deep part of traditional Hindu society and is described in detail in the religious texts (Vedas) that were composed subsequent to the *Rig Veda*.

The *jati* system, which few people outside India understand, is much more complicated, and involves a minimum of forty-six hundred and by some accounts around forty thousand endogamous groups.[32] Each is assigned a particular rank in the *varna* system, but strong and complicated endogamy rules prevent people from most different *jatis* from mixing with each other, even if they are of the same *varna* level. It is also clear that in the past, whole *jati* groups have changed their *varna* ranks. For example, the Gujjar *jati* (from which the state of Gujarat in northwest India takes its name) have a variety of ranks depending on where in India they live, which is likely to reflect the fact that in some regions, Gujjars have successfully made the case to raise the status of their *jati* within the *varna* hierarchy.[33]

How the *varna* and *jati* relate to each other is a much-debated mystery. One hypothesis suggested by the anthropologist Irawati Karve is that thousands of years ago, Indian peoples lived in effectively endogamous tribal groups that did not mix, much like tribal groups in other parts of the world today.[34] Political elites then ensconced themselves at the top of the social system (as priests, kings, and merchants), creating a stratified system in which the tribal groups were incorporated into society in the form of laboring groups that remained at the bottom of society as Shudras and Dalits. The tribal organization was thus fused with the system of social stratification to form early *jatis*, and eventually the *jati* structure percolated up to the higher ranks of society, so that today there are many *jatis* of higher as well as of lower castes. These ancient tribal groups have preserved their distinctiveness through the caste system and endogamy rules.

An alternative hypothesis is that strong endogamy rules are not very old at all. The theory of the caste system is undeniably old, as it is described in the ancient *Law Code of Manu*, a Hindu text composed some hundreds of years after the *Rig Veda*. The *Law Code of Manu* describes in exquisite detail the *varna* system of ranked social stratification, and within it the innumerable *jati* groups. It puts the whole system into a religious framework, justifying its existence as part of the natural order of life. However, revisionist historians, led by the anthropologist Nicholas Dirks, have argued that in fact strong endogamy was not practiced in ancient India, but instead is largely an innovation of British colonialism.[35] Dirks and colleagues showed how, as a way of effectively ruling India, British policy beginning in the eighteenth century was to strengthen the caste system, carving out a natural place within Indian society for British colonialists as a new caste group. To achieve this, the British strengthened the institution of caste in parts of India where it was not very important, and worked to harmonize caste rules across different regions. Given these efforts, Dirks suggested that strong endogamy restrictions as manifested in today's castes might not be as old in practice as they seem.

To understand the extent to which the *jatis* corresponded to real genetic patterns, we examined the degree of differentiation of each *jati* from which we had data with all others based on differences in mutation frequencies.[36] We found that the degree of differentiation was at least three times greater than that among European groups separated by similar geographic distances. This could not be explained by differences in ANI ancestry among groups, or differences in the region within India from which the population came, or differences in social status. Even comparing pairs of groups matched according to these criteria, we found that the degree of genetic differentiation among Indian groups was many times larger than that in Europe.

These findings led us to surmise that many Indian groups today might be the products of population bottlenecks. These occur when relatively small numbers of individuals have many offspring and their descendants too have many offspring and remain genetically isolated from the people who surround them due to social or geographic barriers. Famous population bottlenecks in the history of people of European ancestry include the ones that contributed most of the ancestry of the Finnish population (around two thousand years ago), a large fraction of the ancestry of today's Ashkenazi Jews (around six hundred years ago), and most of the ancestry of religious dissenters such as Hutterites and Amish who eventually migrated to North America (around three hundred years ago). In each case, a high reproductive rate among a small number of individuals caused the rare mutations carried in those individuals to rise in frequency in their descendants.[37]

We looked for the telltale signs of population bottlenecks in India and found them: identical long stretches of sequence between pairs of individuals within the same group. The only possible explanation for such segments is that the two individuals descend from an ancestor in the last few thousand years who carried that DNA segment. What's more, the average size of the shared DNA segments reveals how long ago in the past that shared ancestor lived, as the shared segments break up at a regular rate in each generation through the process of recombination.

The genetic data told a clear story. Around a third of Indian groups experienced population bottlenecks as strong or stronger than the ones that occurred among Finns or Ashkenazi Jews. We later confirmed this finding in an even larger dataset that we collected working with Thangaraj: genetic data from more than 250 *jati* groups spread throughout India.[38]

Many of the population bottlenecks in India were also exceedingly old. One of the most striking we discovered was in the Vysya of the southern Indian state of Andhra Pradesh, a middle caste group of approximately five million people whose population bottleneck we could date (from the size of segments shared between individuals of the same population) to between three thousand and two thousand years ago.

The observation of such a strong population bottleneck among the ancestors of the Vysya was shocking. It meant that after the population bottleneck, the ancestors of the Vysya had maintained strict endogamy, allowing essentially no genetic mixing into their group for thousands of years. Even an average rate of influx into the Vysya of as little as 1 percent per generation would have erased the genetic signal of a population bottleneck. The ancestors of the Vysya did not live in geographic isolation. Instead, they lived cheek by jowl with other groups in a densely populated part of India. Despite proximity to other groups, the endogamy rules and group identity in the Vysya have been so strong that they maintained strict social isolation from their neighbors, and transmitted that culture of social isolation to each and every subsequent generation.

And the Vysya were not unique. A third of the groups we analyzed gave similar signals, implying thousands of groups in India like this. Indeed, it is even possible that we were underestimating the fraction of groups in India affected by strong long-term endogamy. To show a signal, a group needed to have gone through a population bottleneck. Groups that descended from a larger number of founders but nevertheless maintained strict endogamy ever since would go undetected by our statistics. Rather than an invention of colonialism as Dirks suggested, long-term endogamy as embodied in India today in the institution of caste has been overwhelmingly important for millennia.

Learning this feature of Indian history had a strong resonance for me. When I started my work on Indian groups, I came to it as an Ashkenazi Jew, a member of an ancient caste of West Eurasia. I was uncomfortable with my affiliation but did not have a clear sense of what I was uncomfortable about. My work on India crystallized my discomfort. There is no escaping my background as a Jew. I was raised by parents whose highest priority was being open to the secular world, but they themselves had been raised in a deeply religious community and were children of refugees from persecution in Europe that left them with a strong sense of ethnic distinctiveness. When I was growing up, we followed Jewish dietary rules at home—I believe my parents did so in part in the hope that their own families would feel comfortable eating at our house—and I went for nine years to a Jewish school and spent many summers in Jerusalem. From my parents as well as from my grandparents and cousins I imbibed a strong sense of difference—a feeling that our group was special—and a knowledge that I would cause disappointment and embarrassment if I married someone non-Jewish (a conviction that I know also had a powerful effect on my siblings). Of course, my concern about disappointing my family is nothing compared to the shame, isolation, and violence that many expect in India for taking a partner outside their group. And yet my perspective as a Jew made me empathize strongly with all the likely Romeos and Juliets over thousands of years of Indian history whose loves across ethnic lines have been quashed by caste. My Jewish identity also helped me to understand on a visceral level how this institution had successfully perpetuated itself for so long.

What the data were showing us was that the genetic distinctions among *jati* groups within India were in many cases real, thanks to the long-standing history of endogamy in the subcontinent. People tend to think of India, with its more than 1.3 billion people, as having a tremendously large population, and indeed many Indians as well as foreigners see it this way. But genetically, this is an incorrect way to view the situation. The Han Chinese are truly a large population. They have been mixing freely for thousands of years. In contrast, there are few if any Indian groups that are demographically very large, and the degree of genetic differentiation among Indian *jati* groups living side by side in the same village is typically two to three times higher than the genetic differentiation between northern and southern Europeans.[39] The truth is that India is composed of a large number of small populations.

Indian Genetics, History, and Health

The groups of European ancestry that have experienced strong population bottlenecks—Ashkenazi Jews, Finns, Hutterites, Amish, French Canadians of the Saguenay–Lac-St.-Jean region, and others—have been the subject of endless and productive study by medical researchers. Because of their population bottlenecks, rare disease-causing mutations that happened to have been carried in the founder individuals have dramatically increased in frequency. Rare mutations that are innocuous when a person inherits a copy from only one of their parents—they act recessively, which means that two copies are required to cause disease—can be lethal when a person inherits copies from both parents. However, once these mutations increase in frequency due to a population bottleneck, there is an appreciable chance that individuals in the population will inherit the same mutation from both of their parents. For example, in Ashkenazi Jews there is a high incidence of the devastating disease of Tay-Sachs, which causes brain degeneration and death within the first few years of life. One of my first cousins died within months of birth due to an Ashkenazi founder disease called Zellweger syndrome, and one of my mother's first cousins died young of Riley-Day syndrome, or familial dysautonomia, another Ashkenazi founder disease. Hundreds of such diseases have been identified, and the responsible genes have been identified in European founder populations, including Ashkenazi Jews. These findings have led to important biological insights and in a few cases to the development of drugs that counteract the effect of the damaged genes.

India, of course, has far more people who belong to groups that experienced strong bottlenecks, as the country's population is huge, and as around one-third of Indian *jati* groups descend from bottlenecks as strong or stronger than those that occurred in Ashkenazi Jews or Finns. Searches for the genes responsible for disorders in these Indian groups therefore have the potential to identify risk factors for thousands of diseases. Despite the fact that no one has systematically looked, a few such cases are already known. For example, the Vysya are known to have a high rate of prolonged muscle paralysis in response to muscle relaxants given prior to surgery. As a result, clinicians in India know not to give these drugs to people of Vysya ancestry. The condition is due to low levels of the protein butylcholinesterase in some Vysya. Genetic work has shown that this condition is due to a recessively acting mutation that occurs at about 20 percent frequency in the Vysya, a far higher rate than in other Indian groups, presumably because the mutation was carried in one of the Vysya's founders.[40] This frequency is sufficiently high that the mutation occurs in two copies in about 4 percent of the Vysya, causing disastrous reactions for people who carry the mutation and go under anesthesia.

As the Vysya example demonstrates, the history of India presents an important opportunity for biological discovery, as finding genes for rare recessive diseases is cheap with modern genetic technology. All it takes is access to a small number of people in a *jati* group with the disease, whose genomes can then be sequenced. Genetic methods can identify which of the thousands of groups in India have experienced strong population bottlenecks. Local doctors and midwives can identify syndromes that occur at high rates in specific groups. It is surely the case that local doctors, having delivered thousands of babies, will know that certain diseases and malformations occur more frequently in some groups than in others. This is all the information one needs to collect a handful of blood samples for genetic analysis. Once these samples are in hand, the genetic work to find the responsible genes is straightforward.

The opportunities for making a medical difference in India through surveys of rare recessive disease are particularly great because arranged marriage is very common. Much as I find restrictions on marriage discomfiting, arranged marriages are a fact in numerous communities in India—as they are in the ultra-Orthodox Jewish community. A number of my own first cousins in the Ashkenazi Jewish Orthodox community have found their spouses that way. In this religious community, a genetic testing organization founded by Rabbi Josef Ekstein in 1983, after he lost four of his own children to Tay-Sachs, has driven many recessive diseases almost to extinction.[41] In many Orthodox religious high schools in the United States and Israel, nearly all teenagers are tested for whether they are carriers of the handful of rare recessive disease-causing mutations that are common in the Ashkenazi Jewish community. If they are carriers, they are never introduced by matchmakers to other teenagers carrying the same mutation. There is every opportunity to do the same in India, but instead of affecting a few hundred thousand people, in India the approach could have an impact on hundreds of millions.

A Tale of Two Subcontinents:
The Parallel History of India and Europe

Up until 2016, the genetic studies of Indian groups focused on the ANI and the ASI: the two populations that mixed in different proportions to produce the great diversity of endogamous groups still living in India today.

But this changed in 2016, when several laboratories, including mine, published the first genome-wide ancient DNA from some of the world's earliest farmers, people who lived between eleven thousand and eight thousand years ago in present-day Israel, Jordan, Anatolia, and Iran.[42] When we studied how these early farmers of the Near East were related to people living today, we found that present-day Europeans have strong genetic affinity to early farmers from Anatolia, consistent with a migration of Anatolian farmers into Europe after nine thousand years ago. Present-day people from India have a strong affinity to ancient Iranian farmers, suggesting that the expansion of Near Eastern farming eastward to the Indus Valley after nine thousand years ago had as important an impact on the population of India.[43] But our studies also revealed that present-day people in India have strong genetic affinities to ancient steppe pastoralists. How could the genetic evidence of an impact of an Iranian farming expansion on the population of India be reconciled with the evidence of steppe expansions? The situation was reminiscent of what we had found a couple of years before in Europe, where today's populations are a mixture not just of indigenous hunter-gatherers and migrant farmers, but also of a third major group with an origin in the steppe.

To gain some insight, Iosif Lazaridis in my laboratory wrote down mathematical models for present-day Indian groups as mixtures of populations related to Little Andaman Islanders, ancient Iranian farmers, and ancient steppe peoples. What he found is that almost every group in India has ancestry from all three populations.[44] Nick Patterson then combined the data from almost 150 present-day Indian groups to come up with a unified model that allowed him to obtain precise estimates of the contribution of these three ancestral populations to present-day Indians.

When Patterson inferred what would have been expected for a population of entirely ANI ancestry—one with no Andamanese-related ancestry—he determined that they would be a mixed population of Iranian farmer–related ancestry and steppe pastoralist–related ancestry. But when he inferred what would have been expected for a population of entirely ASI ancestry—one with no Yamnaya-related ancestry—he found that they too must have had substantial Iranian farmer–related ancestry (the rest being Little Andamanese–related).

This was a great surprise. Our finding that both the ANI and ASI had large amounts of Iranian-related ancestry meant that we had been wrong in our original presumption that one of the two major ancestral populations of the Indian Cline had no West Eurasian ancestry. Instead, people descended from Iranian farmers made a major impact on India twice, admixing both into the ANI and the ASI.

Patterson proposed a major revision to our working model for deep Indian history.[45] The ANI were a mixture of about 50 percent steppe ancestry related distantly to the Yamnaya, and 50 percent Iranian farmer–related ancestry from the groups the steppe people

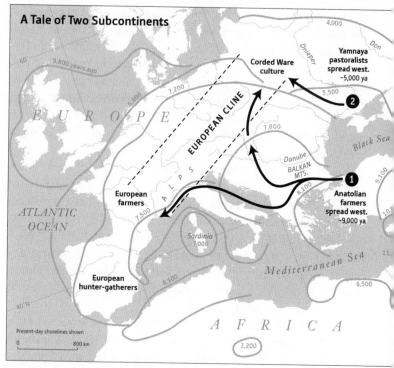

Figure 18. Both South Asia and Europe were affected by two successive migrations. The first migration was from the Near East after around nine thousand years ago (1), which brought farmers who mixed with local hunter-gatherers. The second migration was from the steppe after around

encountered as they expanded south. The ASI were also mixed, a fusion of a population descended from earlier farmers expanding out of Iran (around 25 percent of their ancestry), and previously established local hunter-gatherers of South Asia (around 75 percent of their ancestry). So the ASI were not likely to have been the previously established hunter-gatherer population of India, and instead may have been the people responsible for spreading Near Eastern agriculture across South Asia. Based on the high correlation of ASI ancestry to Dravidian languages, it seems likely that the formation of the ASI was the process that spread Dravidian languages as well.

These results reveal a remarkably parallel tale of the prehistories

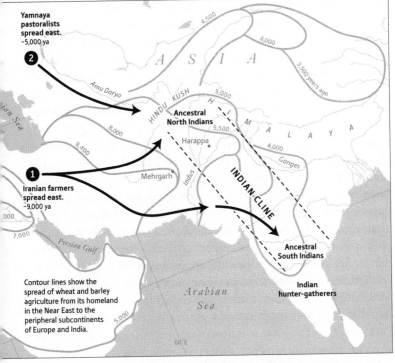

Contour lines show the spread of wheat and barley agriculture from its homeland in the Near East to the peripheral subcontinents of Europe and India.

five thousand years ago (2), which brought pastoralists who probably spoke Indo-European languages, who then mixed with the local farmers they encountered along the way. Mixtures of these mixed groups then formed two gradients of ancestry: one in Europe, and one in India.

of two similarly sized subcontinents of Eurasia—Europe and India. In both regions, farmers migrating from the core region of the Near East after nine thousand years ago—in Europe from Anatolia, and in India from Iran—brought a transformative new technology, and interbred with the previously established hunter-gatherer populations to form new mixed groups between nine thousand and four thousand years ago. Both subcontinents were then also affected by a second later major migration with an origin in the steppe, in which Yamnaya pastoralists speaking an Indo-European language mixed with the previously established farming population they encountered along the way, in Europe forming the peoples associated with the Corded Ware culture, and in India eventually forming the ANI. These populations of mixed steppe and farmer ancestry then mixed with the previously established farmers of their respective regions, forming the gradients of mixture we see in both subcontinents today.

The Yamnaya—who the genetic data show were closely related to the source of the steppe ancestry in both India and Europe—are obvious candidates for spreading Indo-European languages to both these subcontinents of Eurasia. Remarkably, Patterson's analysis of population history in India provided an additional line of evidence for this. His model of the Indian Cline was based on the idea of a simple mixture of two ancestral populations, the ANI and ASI. But when he looked harder and tested each of the Indian Cline groups in turn for whether it fit this model, he found that there were six groups that did not fit in the sense of having a higher ratio of steppe-related to Iranian farmer–related ancestry than was expected from this model. All six of these groups are in the Brahmin *varna*—with a traditional role in society as priests and custodians of the ancient texts written in the Indo-European Sanskrit language—despite the fact that Brahmins made up only about 10 percent of the groups Patterson tested. A natural explanation for this was that the ANI were not a homogeneous population when they mixed with the ASI, but instead contained socially distinct subgroups with characteristic ratios of steppe to Iranian-related ancestry. The people who were custodians of Indo-European language and culture were the ones with relatively more steppe ancestry, and because of the extraordinary strength of the caste system in preserving ancestry and social roles over generations, the ancient substructure in the ANI is evident in some of today's Brahmins even after thousands of years. This finding provides yet another line of evidence for the steppe hypothesis, showing that not just Indo-European languages, but also Indo-European culture as reflected in the religion preserved over thousands of years by Brahmin priests, was likely spread by peoples whose ancestors originated in the steppe.

The picture of population movements in India is still far less crisp than our picture of Europe because of the lack of ancient DNA from South Asia. An outstanding mystery is the ancestry of the peoples of the Indus Valley Civilization, who were spread across the Indus Valley and parts of northern India between forty-five hundred to thirty-eight hundred years ago, and were at the crossroads of all these great ancient movements of people. We have yet to obtain ancient DNA from the people of the Indus Valley Civilization, but multiple research groups, including mine, are pursuing this as a goal. At a lab meeting in 2015, the analysts in our group went around the table placing bets on the likely genetic ancestry of the Indus Valley Civilization people, and the bets were wildly different. At the moment, three very different possibilities are still on the table. One is that Indus Valley Civilization people were largely unmixed descendants of the first Iranian-related farmers of the region, and spoke an early Dravidian language. A second possibility is that they were the ASI—already a mix of people related to Iranian farmers and South Asian hunter-gatherers—and if so they would also probably have spoken a Dravidian language. A third possibility is that they were the ANI, already mixed between steppe and Iranian farmer–related ancestry, and thus would instead likely have spoken an Indo-European language. These scenarios have very different implications, but with ancient DNA, this and other great mysteries of the Indian past will soon be resolved.

Migrations to the Americas

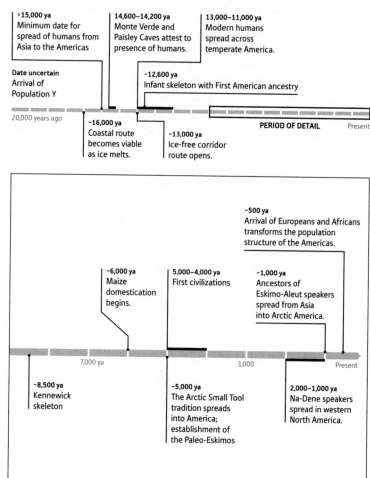

>15,000 ya
Minimum date for spread of humans from Asia to the Americas

14,600–14,200 ya
Monte Verde and Paisley Caves attest to presence of humans.

13,000–11,000 ya
Modern humans spread across temperate America.

Date uncertain
Arrival of Population Y

~12,600 ya
Infant skeleton with First American ancestry

20,000 years ago

~16,000 ya
Coastal route becomes viable as ice melts.

~13,000 ya
Ice-free corridor route opens.

PERIOD OF DETAIL   Present

~500 ya
Arrival of Europeans and Africans transforms the population structure of the Americas.

~6,000 ya
Maize domestication begins.

5,000–4,000 ya
First civilizations

~1,000 ya
Ancestors of Eskimo-Aleut speakers spread from Asia into Arctic America.

7,000 ya   3,000   Present

~8,500 ya
Kennewick skeleton

~5,000 ya
The Arctic Small Tool tradition spreads into America; establishment of the Paleo-Eskimos

2,000–1,000 ya
Na-Dene speakers spread in western North America.

9,000 years ago – present

7

In Search of Native American Ancestors

Origins Stories

In an origins story of the Suruí tribe of Amazonia, the god Palop first made his brother, Palop Leregu, and then created humans. Palop gave the Native American tribes hammocks and ornaments and told them to tattoo their bodies and pierce their lips, but he did not give any of these things to the whites. Palop created languages, one for each group, and scattered the groups across the earth.[1]

This origins story was documented by an anthropologist working to understand Suruí culture, and, like origins stories the world over, it is viewed by scholars as fictional, of interest because of what it reveals about a society. But we scientists too have origins stories. We like to think these are superior because they are tested by the scientific method against a range of evidence. But some humility is in order. In 2012, I led a study that claimed that all Native Americans from Mesoamerica southward—including the Suruí—derived all of their ancestry from a single population, one that moved south of the ice sheets sometime after fifteen thousand years ago.[2] I was so confident of this theory, which fit with the consensus derived from archaeology, that I used the term "First American" to signal that the lineage we had highlighted was a founding lineage. Three years later, I found out I was wrong. The Suruí and some of their neighbors in Amazonia harbor some ancestry from a different founding popula-

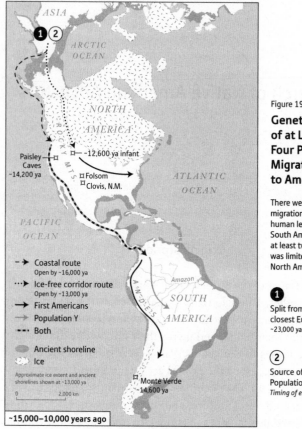

Figure 19

Genetic Evidence of at Least Four Prehistoric Migrations to America

There were at least two migrations that left a human legacy as far as South America (left) and at least two whose impact was limited to northern North America (right).

1 Split from closest Eurasians ~23,000 ya

2 Source of Population Y *Timing of entry unknown*

Paisley Caves ~14,200 ya

~12,600 ya infant

Folsom

Clovis, N.M.

Monte Verde ~14,600 ya

- → Coastal route
  Open by ~16,000 ya
···▶ Ice-free corridor route
  Open by ~13,000 ya
→ First Americans
→ Population Y
--- Both

Ancient shoreline

Ice

Approximate ice extent and ancient shorelines shown at ~13,000 ya

0   2,000 km

~15,000–10,000 years ago

tion of the Americas, whose ancestors arrived at a time and along a route we still do not understand.[3]

If there is anything that scholars studying the history of humans in the Americas agree on, it is that the span of human occupation of the New World has been the blink of an eye relative to the extraordinary length of the human occupation of Africa and Eurasia. The reason for humans' late arrival to America lies in the geographical barriers that separate the continent from Eurasia: vast stretches of cold, harsh, and unproductive landscapes in Siberia, and oceans to the east and west. It took until the last ice age for Siberia's northeastern corner to be visited by people with the skills and technology needed to

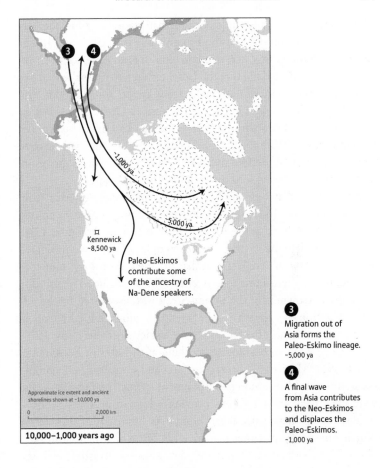

3
Migration out of
Asia forms the
Paleo-Eskimo lineage.
~5,000 ya

4
A final wave
from Asia contributes
to the Neo-Eskimos
and displaces the
Paleo-Eskimos.
~1,000 ya

Kennewick
~8,500 ya

Paleo-Eskimos
contribute some
of the ancestry of
Na-Dene speakers.

Approximate ice extent and ancient
shorelines shown at ~10,000 ya

0   2,000 km

10,000–1,000 years ago

survive there at a time when sea levels were low enough for a land
bridge to emerge in what is now the Bering Strait region, enabling
them to walk across to Alaska. Once there, the migrants were able to
survive, but they still could not have traveled south, at least by land,
as they were blocked by a wall of glacial ice formed by the joining
together of kilometer-thick ice sheets that buried Canada.

How were the Americas first peopled? Until two decades ago, the
prevailing hypothesis was that the gates of the American Eden only
opened after around thirteen thousand years ago. Evidence from
plant and animal remains and the radiocarbon dating of glacial features indicate that by this time, the ice sheets had melted enough to

allow a gap to open, and sufficient time had passed to allow the barren rocks, mud, and glacial runoff to give way to vegetation.[4] In scientific storytelling, this "ice-free corridor" was an American version of the channel of dry land that the Israelites used to cross the Red Sea in the biblical account of the Exodus from Egypt. The migrants who passed through emerged into North America's Great Plains. Before them was a land filled with massive game that had never before met human hunters. Within a thousand years, the humans had reached Tierra del Fuego at the foot of South America, feasting on the bison, mammoths, and mastodons that roamed the landscape.

The notion that humans first reached an empty America from Asia—an idea that today is still the overwhelming consensus among scholars—dates back to the Jesuit naturalist José de Acosta in 1590, who, finding it unlikely that ancient peoples could have navigated across a great ocean, conjectured that the New World was joined to the Old in the then-unmapped Arctic.[5] This idea gained plausibility when the narrowness of the Bering Strait was discovered by the circumnavigation of Captain Cook. Scientific evidence for humans in temperate America at the tail end of the last ice age came in the 1920s and 1930s, when archaeologists working at the sites of Folsom and Clovis, New Mexico, discovered artifacts and stone tools—including spear tips mixed in among the bones of extinct mammoths—that were effectively smoking guns proving a human presence. Clovis-style spear tips have since been found over hundreds of sites across North America, sometimes embedded in bison and mammoth skeletons. Their similar style over vast distances—contrasting with the regional variation in stone toolmaking styles of the cultures that followed—is what one might expect for an expansion that occurred fast (as the people were moving into a human vacuum). The available evidence suggests that the Clovis culture appeared in the archaeological record around the time of the geologically attested opening of the ice-free corridor, so everything seemed to fit. It seemed natural to think that people practicing the Clovis culture were the first humans south of the ice sheets, and were also the ancestors of all of today's Native Americans.

This "Clovis First" model, in which the makers of the Clovis culture emerged from the ice-free corridor and proceeded to people an empty continent, became the standard model of American prehistory. It fostered skepticism among archaeologists regarding claims of pre-Clovis sites.[6] It influenced linguists who claimed to find a common origin for a large number of diverse Native American languages.[7] The mitochondrial DNA data available at the time was also consistent with the great majority of the ancestry of present-day Native Americans deriving from a radiation from a single source, although with such data alone it was not possible to determine whether that radiation occurred at the time of Clovis or before.[8]

A major blow to the idea that Clovis groups were the first Americans came in 1997. That year marked the publication of the results of excavations at the site of Monte Verde in Chile, which contains butchered mastodon bones, wooden remains of structures, knotted string, ancient hearths, and stone tools with no stylistic similarities to the Clovis remains from North America.[9] The radiocarbon dates of Monte Verde indicated that some of the artifacts there dated to around fourteen thousand years ago, definitively before the ice-free corridor had opened thousands of kilometers to the north. A group of skeptical archaeologists who had previously shot down many pre-Clovis claims visited the site that same year, and though they arrived doubting that the site could be that old, they left convinced. Their acceptance of Monte Verde was followed by the acceptance of finds elsewhere that also pointed to a pre-ice-free corridor and a pre-Clovis human presence in the Americas. Nearly as strong a case for a pre-ice-free corridor occupation has been made at the Paisley Caves in Oregon in the northwestern United States, where ancient feces in undisturbed soil layers have also been dated to around fourteen thousand years ago, and have yielded human mitochondrial DNA sequences.[10]

How could humans have gotten south of the ice sheets before the ice-free corridor was open? During the peak of the ice age, glaciers projected right into the sea, creating a barrier more than a thousand kilometers in length along the western seaboard of Canada. But in the 1990s, geologists and archaeologists, reconstructing the timing of the ice retreat, realized that portions of the coast were ice-free after sixteen thousand years ago. There are no known archaeological sites along the coast from this period, as sea levels have risen more than a hundred meters since the ice age, submerging any archaeological sites that might have once hugged the shoreline. The absence of archaeological evidence for human occupation along the coast in this period is therefore not evidence that there was no such occupation in the past. If the coastal route hypothesis is right, humans could have walked at that time or later (but still in time to reach Monte Verde) along ice-free stretches of the coastline, possibly bypassing ice-covered sections with boats or rafts, and arriving south of the ice millennia before the interior ice-free corridor opened.

Ancient DNA studies have also now made it clear just how wrong the Clovis First idea is—how it misses a whole deep branch of Native American population history. In 2014, Eske Willerslev and his colleagues published whole-genome data from the remains of an infant excavated in Montana whose archaeological context assigned him to the Clovis culture and whose radiocarbon age was a bit after thirteen thousand years ago.[11] Their analysis showed that this infant was definitely from the same ancestral population as many Native Americans, but his genetic data also showed that by the time he lived, a deep split among Native American populations had already developed. The remains from the Clovis infant were on one side of that split: the side that contributed the lion's share of ancestry to all Native American populations in Mesoamerica and South America today. The other side of the split includes Native American peoples who today live in eastern and central Canada. The only way this could have happened is if there had been a population that lived before Clovis and that gave rise to major Native American lineages.

Mistrust of Western Science

Ancient DNA studies such as the one of the Clovis infant have the potential to resolve controversies about Native American population history. But such studies have resonances for present-day descendants of those populations that are not entirely positive. That is because the last five hundred years have witnessed repeated cases in which people of European ancestry have exploited the indigenous peoples of the Americas using the toolkit of Western science. This has engendered distrust between some Native American groups and the scholarly community—a distrust that makes carrying out genetic studies challenging.

After the arrival of Europeans in the Americas in 1492, Native American populations and cultures collapsed under the pressure of European diseases, military campaigns, and an economic and political regime set on exploiting the riches of the continent and converting its inhabitants to Christianity. History is written by the victors, and the rewriting of the past after the European conquests has been particularly complete in the Americas, as there was no written language except in Mesoamerica prior to the arrival of Europeans. In Mexico, the Spanish burned indigenous books, and so most Native American writing literally went up in flames. The oral traditions suffered too. Language change, religious conversion, and discrimination against indigenous ways led Native American culture to be relegated to a lower status than European culture.

Modern genomics offers an unexpected way to recover the past. African Americans—another population that has had its history stolen as its ancestors descend from people kidnapped into slavery from Africa—are at the forefront of trying to use genetics to trace roots. But if individual Native Americans often express a great interest in their genetic history, tribal councils have sometimes been hostile. A common concern is that genetic studies of Native American history are yet another example of Europeans trying to "enlighten" them. Past attempts to do so—for example, by conversion to Christianity or education in Western culture—have led to the dissolution of Native American culture. There is also an awareness that some scientists have studied Native Americans to learn about questions of interest primarily to non–Native Americans, without paying attention to the interests of Native Americans themselves.

One of the first strong responses to genetic studies of Native Americans came from the Karitiana of Amazonia. In 1996, physicians collected blood from the Karitiana, promising participants improved access to health care, which never came. Distressed by this experience, the Karitiana were at the forefront of objections to the inclusion of their samples in an international study of human genetic diversity—the Human Genome Diversity Project—and were instrumental in preventing that entire project from being funded. Ironically, DNA samples from the Karitiana have been used more than those of any other single Native American population in subsequent studies that have analyzed how Native Americans are related to other groups. The Karitiana DNA samples that have been widely studied are not from the disputed set from 1996. Instead, they are from a collection carried out in 1987 in which participants were informed about the goals of the study and told that their involvement was voluntary.[12] However, the Karitiana people's later experience of exploitation has put a cloud over DNA studies in this population.

Another strong response to genetic research on Native Americans came from the Havasupai, who live in the canyonlands of the U.S. Southwest. Blood from the Havasupai was sampled in 1989 by researchers at Arizona State University who were trying to understand the tribe's high risk for type 2 diabetes. The participants gave written consent to participate in a "study [of] the causes of behavioral/medical disorders," and the language of the consent forms gave the researchers latitude to take a very broad view of what the consent meant. The researchers then shared the samples with many other scientists who used them to study topics ranging from schizophrenia to the Havasupai's prehistory. Representatives of the Havasupai argued that the samples were being used for a purpose different from the one to which its members understood they had agreed—that is, even if the fine print of the forms said one thing, it was clear to them when the samples were collected that the study was supposed to focus on diabetes. This dispute led to a lawsuit, the return of the samples, and an agreement by the university to pay $700,000 in compensation.[13]

The hostility to genetic research has even entered into tribal law. In 2002, the Navajo—who along with many other Native American tribes are by treaty partly politically independent of the United States—passed a Moratorium on Genetic Research, forbidding participation of Navajo tribal members in genetic studies, whether of disease risk factors or population history. A summary of this moratorium can be found in a document prepared by the Navajo Nation, outlining points for university researchers to take into account when considering a research project. The document reads: "Human genome testing is strictly prohibited by the Tribe. Navajos were created by Changing Woman; therefore they know where they came from."[14]

I became aware of the Navajo moratorium in 2012, while I was in the final stages of preparing a manuscript on genetic variation among diverse Native Americans. After receiving favorable reviews of our manuscript, I asked each researcher who contributed samples to double-check whether the informed consent associated with the samples was consistent with studies of population history and to confirm that they themselves stood behind the inclusion of their samples in our study. This led to withdrawal of three populations from the study, including the Navajo. All three populations were from the United States, reflecting the anxiety that has seized U.S. genetic researchers about genetic studies of Native Americans. At a workshop on genetic studies of Native Americans that I attended in 2013, multiple researchers stood up from the audience to say that the responses of the Karitiana, Havasupai, Navajo, and others had made them too wary to do any research on Native Americans (including disease research).

Scientists interested in studying genetic variation in Native American populations feel frustrated with this situation. I understand something of the devastation that the coming of Europeans and Africans to the Americas wrought on Native American populations, and its effects are also evident everywhere in the data I and my colleagues analyze. But I am not aware of any cases in which research in molecular biology including genetics—a field that has arisen almost entirely since the end of the Second World War—has caused major harm to historically persecuted groups. Of course, there have been well-documented cases of the use of biological material in ways that may not have been appreciated by the people from whom it was taken, not just in Native Americans. For example, the cervical cancer tumor cells of Henrietta Lacks, an African American woman from Baltimore, were distributed after her death, without her consent and without the knowledge of her family, to thousands of laboratories around the world, where they have become a mainstay of cancer research.[15] But overall there is an argument to be made that modern studies of DNA variation—not just in Native Americans, but also in many other groups including the San of southern Africa, Jews, the Roma of Europe, and tribal or caste groups from South Asia—are a force for good, contributing to the understanding and treatment of disease in these populations, and breaking down fixed ideas of race that have been used to justify discrimination. I wonder if the distrust that has emerged among some Native Americans might be, in the balance, doing Native Americans substantial harm. I wonder whether as a geneticist I have a responsibility to do more than just respect the wishes of those who do not wish to participate in genetic research, but instead should make a respectful but strong case for the value of such research.

The withdrawal of Navajo samples from our study was distressing, since they were among those with the very best documentation of informed consent. The researcher who shared the samples with us had collected them personally in 1993 as part of a "DNA day" that he had organized at Diné College on Navajo lands, so there was no ambiguity involved in the handoff of samples along a human chain. During the workshop, he asked participants if they wished to donate their samples for the explicit purpose of broad studies of population history—specifically for studies that "give prominence to the idea that all peoples of the world are closely related and emphasize the unity of human origins"—and members of the Navajo tribe who wished to participate signed a form indicating that they did. Yet these individuals' personal decisions to participate in the study were overruled by the tribal council's moratorium nine years later.

Should we have respected the wishes of the college students who donated the samples, or the later decision of the tribal council? In the instance, we avoided the issue, acceding to the request of the researcher, who was so concerned that he asked us not to include the samples in the study. I was never comfortable with this decision. I felt that including the samples would best respect the wishes of the individuals who chose to donate their DNA for studies of their history. But I recognize that different cultures have different perspectives. There is a movement among some Native American ethicists and community leaders to argue that any research that has as its subject a tribe should only be considered acceptable if there is community consultation, not just informed individual consent.[16] These concerns prompted some international studies of human genetic variation to carry out community consultation in addition to individual informed consent before including samples.[17] The very few researchers studying Native American genetic diversity almost all now consult with tribal authorities to obtain feedback on study design—and sometimes to obtain explicit community consent—even if doing so is not legally required.

There is a general issue here about the ethical responsibilities of genetic research. When I examine an individual's genome, I learn not only about the genome of the individual, but also about those of his or her family, and ancestors. I also learn about other members of the community—other descendants of those same ancestors. What are my responsibilities here? What do I owe not only to close relatives of the individual I study, but also to other more distantly related members of their family, to their population, and to our species as a whole? An extreme position that everyone needs to be consulted would make scientific progress in human genetics (including genetic medicine) nearly impossible. There would not be enough time for scientists in modest-sized laboratories like mine to talk with every tribal group that might be interested in the work.

My own perspective is that we need as a scientific community to arrive at a middle ground, an approach that does not require obtaining permission from every possible interested group or tribe. On the other hand, given the well-founded concerns of tribal communities in North America, which have developed as a result of a persistent history of exploitation, we scientists should aspire to carry out meaningful outreach when we study Native American population history to ensure that any manuscripts we write are sensitive to indigenous perspectives. The details of how to achieve such consultation need to be worked out, and it seems to me that there will never be a solution that everyone will find comfortable. But we need to try to make progress beyond the situation we are facing right now, in which many researchers are reluctant to undertake any studies of Native American genetic variation for fear of criticism, and because of the extraordinary time commitment that would be required in order to accomplish all the consultations that some tribal representatives and scholars have recommended. This has had the effect of putting research into genetic variation among Native Americans into a deep chill—with far less research in this area going on than anyone but the people most hostile to scientific research would like.

Disputes over Bones

Ancient DNA studies of population history are mostly not as fraught as studies of present-day people. However, in 1990, the U.S. Congress passed the Native American Graves Protection and Repatriation Act (NAGPRA), which requires institutions that receive U.S. funding to contact Native American tribes and offer to return cultural artifacts, including bones that are from groups to which Native Americans can prove a biological or cultural connection. This has meant that Native American remains are being returned to Native American tribes and the opportunity to carry out ancient DNA analysis on many of the samples is disappearing. NAGPRA has had its greatest impact on archaeological remains dating to within the last thousand years, for which a relatively strong case can be made for cultural connections with living Native American tribes. The case for cultural connection is harder to make for very old remains, such as the approximately eighty-five-hundred-year-old Kennewick Man found on U.S. lands in Washington State in 1996.

Kennewick Man's skeleton was initially slated for return to five Native American tribes that claimed him as an ancestor, but was made available for scientific study instead after courts found that there was no good scientific evidence that he was Native American under the rules of NAGPRA. To win their case, the scientists who challenged the tribal claims pointed to analyses of skeletal morphology that suggested that his skeleton was closer to Pacific Rim Asian and Pacific islander populations than to present-day Native Americans.[18] In 2015, though, Eske Willerslev and his colleagues extracted and studied ancient DNA from Kennewick Man, which showed that these conclusions from the morphological studies were wrong.[19] Kennewick Man is in fact derived from the same broad ancestral population as most other Native Americans.

Ancient DNA trumps morphological analysis whenever it is possible to compare the two types of data. The reason is simple. Morphological studies of skeletons can only examine a handful of traits that are variable among individuals, and thus can usually support only uncertain population assignment. In contrast, genetic analyses of tens of thousands of independent positions allow exact population assignment. Thus, the characterization of the ancestry of a single sample (like Kennewick Man) based on a small number of morphological traits cannot convincingly distinguish between Native American and Pacific Rim ancestry. Genetic data can.

While the ancient DNA study produced clear proof of the Native American ancestry of Kennewick Man, it was not so clear whether he bears a particularly strong relationship to the Washington State Native American populations that made claims on his remains. The paper reporting the Kennewick Man genome sampled DNA from the Colville tribe, one of the five tribes staking a claim of relationship to him, and argued that the data were consistent with a direct link. However, the Colville was the only tribe from the lower forty-eight states of the United States that the scientists analyzed, and a close look at the details of the paper provides no compelling case that Kennewick Man is more closely related to the Colville tribe than he is to Native Americans as far away as South America.[20] The Colville data are also not available to the scientific community for independent analysis—they were not provided to my group on request despite the fact that the journal in which they were published requires sharing of data as a condition for publication.

Wishful interpretation of genetic data is not limited to Kennewick Man. In 2017, a study of an approximately 10,300-year-old skeleton excavated from an island off the Pacific coast of present-day Canada, claimed evidence for an unbroken presence of a lineage of Native Americans in the same region from his time until to the present day.[21] But an examination of the analyses presented in the paper showed that this individual, too, was no more closely related to local people than to Native Americans in South America.

These are just two examples of how the ancient DNA literature is beginning to fill up with unsubstantiated claims of direct ancestral links between ancient skeletons and groups living today, a problem that is not limited to the Americas. Scientists working with indigenous people have an incentive to make such claims, as claims like this are often welcomed by local groups, and open the door to sampling. The normal scientific process, in which scientists point out claims that are not compellingly supported by data, is also not working as it should. A concern is that when members of groups are directly engaged in scientific investigation of their own history, people's wish that certain things should be true often colors presentation of the findings. And scientists not involved in the work are often too anxious about repercussions to point out problems.

The Kennewick Man case was contentious and played itself out in court, engendering hostility between academics and Native American tribes. It has had consequences for scientists interested in Native American population history, and it has made such research far more difficult. From my experience interacting with archaeologists, anthropologists, and museum directors who focus on Native American prehistory, it is clear to me that many feel a deep sense of loss about the return of collections of scientifically important bones, and wish to keep them in the possession of museums while acknowledging the dubious ways in which many of these collections were assembled in the course of U.S. expropriation of Native American lands.[22] Balanced against this is the sense of loss that many Native Americans feel about having ancestors' remains disturbed. To navigate these competing interests and the law, many museums employ "NAGPRA officers" whose job it is to identify cultural and skeletal remains that can be associated with particular Native American tribes and to reach out to representatives of those tribes in order to return the items. But while the NAGPRA officers with whom I have interacted are dedicated to fulfilling the letter of the law and do so professionally, they are also careful to not go beyond it. They feel distressed when, as in the case of Kennewick Man, remains are returned to tribes without the evidence of biological or cultural connection that NAGPRA regulations require.

One geneticist who is breaking new ground in this area is Eske Willerslev. Not only with the Kennewick sample but also with other indigenous skeletal remains from which he has assembled DNA, Willerslev has won the cooperation of indigenous communities in a way that is innovative and brilliant—even if it is not making everyone in the archaeological and museum community happy. He has realized that there can be shared interests between indigenous communities and geneticists because DNA studies can empower tribes to stake claims on remains. This happened in the case of the genome sequences extracted from an approximately one-hundred-year-old Australian Aboriginal hair sample,[23] the almost thirteen-thousand-year-old Clovis skeleton,[24] and the approximately eighty-five-hundred-year-old Kennewick skeleton.[25] In all three cases, Willerslev approached tribes directly after obtaining DNA, instead of engaging them through an institutionally run process such as the ones that have been set up through NAGPRA.

Although many in the archaeological community have been concerned about Willerslev's approach of engaging tribes outside the formal institutional process, he has been successful in several ways. In Australia, his engagement with Aboriginal groups in the context of his work on the hundred-year-old hair sample generated goodwill and opened the door to a much more ambitious study of present-day Aboriginal populations published in 2016 by him and colleagues.[26] Similarly, in the United States, Willerslev's engagement with indigenous groups in the Clovis and Kennewick cases has helped generate goodwill and encouraged tribes to support ancient DNA analysis of other remains.

A remarkable example of this progress is provided by remains found in Spirit Cave in Utah. In 2000, the U.S. Bureau of Land Management decided against returning these almost eleven-thousand-year-old remains to the Fallon Paiute-Shoshone tribe that requested them. The bureau's basis for the decision was that there was no evidence of biological or cultural connection to that tribe. The tribe then sued, putting the remains into a legal limbo that allowed them to be investigated only for the purpose of studying their ancestry to determine whether they indeed might have a biological connection to the Fallon Paiute-Shoshone. In October 2015, after publication of the Kennewick paper, Willerslev was given access to the remains for ancient DNA analysis, and around a year later he delivered to the bureau a technical report showing that the individual had ancestry that was entirely from the same deep lineage as present-day Native Americans. On the basis of this report, the bureau decided to return the bones to the tribe.[27]

This decision confused the NAGPRA officer I corresponded with about it, who noted that the interpretation was beyond the letter of the NAGPRA law, which required documentation of a connection to the Fallon Paiute-Shoshone more than to other groups, which Willerslev had apparently not demonstrated. But when I talked with Willerslev about returning samples to tribes, his view was that the letter of the NAGPRA law was not so important and that the community standard was changing even if the law had not yet caught up. In an article in the scientific journal *Nature* about the decision to return the Spirit Cave remains, the anthropologist Dennis O'Rourke was quoted as saying that the case set an example for how Native American groups could be engaged in using genetics to determine which remains to study and rebury. The anthropologist Kim TallBear pointed out how the Spirit Cave example showed that the relationship between tribes and scientists need not be antagonistic: "Tribes do not like having a scientific world view politically shoved down their throat . . . but there is interest in the science."[28]

Willerslev's realization that ancient DNA data provide a type of evidence that can be used to establish claims on unaffiliated remains held in museum collections offers an unexpected opportunity to begin to break the logjam of poor relations that has built up between scholars and indigenous communities.

There is also a second great area of unrealized common cause between Native Americans and geneticists—the potential to use ancient DNA to measure the sizes of populations that existed prior to 1492 by looking at variation within the genome of ancient samples. This is a critical issue for Native Americans, as there is evidence for about a tenfold collapse in population size in the Americas following the arrival of Europeans and the waves of epidemic disease that Europeans brought, leading to the dissolution of previously established complex societies. The relatively small population sizes that European colonialists encountered when they arrived in the Americas were used to provide moral justification for the annexation of Native American lands. The European colonialists had an interest in minimizing the estimates of Native American population sizes, of claiming that there were few if any civilizations or sophisticated populations in the Americas before Europeans came.[29]

I hope that as the consequences of the genome revolution are more broadly realized, indigenous people will increasingly recognize how DNA can become a tool to connect present-day Native American people to their roots and to each other. This will not solve all the concerns that Native American ethicists and community leaders have articulated, but it may serve to reduce antagonism and promote greater understanding and even collaboration in the future.

The Genetic Evidence of the First Americans

The first genome-scale study of Native American population history came in 2012, when my laboratory published data on fifty-two diverse populations. A major limitation of the study was that we had no samples at all from the lower forty-eight states of the United States because of anxieties about genetic research on Native Americans. Nevertheless, the study sampled Native American diversity in much of the rest of the hemisphere, and provided new insights about the past.[30]

Most of the individuals we studied derived small fractions of their genomes from African or European ancestors in the last five hundred years, reflecting the profound upheavals that have occurred since the arrival of European colonists. We carried out many analyses on individuals with no evidence of such mixture, but for some populations, especially in Canada, all the individuals we sampled had at least some non–Native American ancestry. Because we wanted to include these populations, we used a technique that allowed us to identify which sections of people's genomes were of European or African origin. We did this by searching for extended genomic stretches in which individuals carried genetic variants at high frequency in Africans and Europeans but at low frequency in Native Americans. Masking out these sections of the genome helped us to peel back the history of five hundred years of admixture in the Americas to understand something about what the structure of Native American population relationships was like before European contact.

We compared all possible pairs of Native American populations using the Four Population Test. We used this test to evaluate whether Eurasian populations—for instance, Han Chinese—shared more genetic mutations with one Native American population or another, testing all possible pairs of populations. For forty-seven of the fifty-two populations, we could not detect differences in their relatedness to Asians. This suggested to us that the vast majority of Native Americans today, including all those from Mexico southward as well as populations from eastern Canada, descend from a single common lineage. (Five remaining populations, all from the Arctic or from the Pacific Northwest coast of Alaska and Canada, also had evidence of ancestry from different lineages.) Thus the extraordinary physical differences among Native American groups today are due to evolution since splitting from a common ancestral population, not to immigration from different sources in Eurasia. We called this common ancestral population the "First Americans."

We hypothesized that the "First American" lineage that we had characterized represented the descendants of the first people to spread south of the ice sheets, whether via an ice-free corridor or along a coastal route. Genomic studies so far have not been able to determine how small this group was or how many generations it wandered. But whatever happened, we were arguing that this was a pioneer population of limited size that moved into a human vacuum, expanding dramatically wherever it arrived.

The genetic data provide support for the correctness of this hypothesis in its broad outlines. As we applied the Four Population Test time and again, it became clear to us that the great majority of Native Americans, from populations in northern North America down to southern South America, can be broadly described as branches of one tree, forming a sharp contrast to patterns of population relationships in Eurasia. Most populations branched cleanly off the central trunk with little subsequent mixture. The splits proceeded roughly in a north-to-south direction, consistent with the idea that as populations traveled south, groups peeled off and settled, remaining in approximately the same place ever since. The most striking exception to this pattern was the less than thirteen-thousand-year-old infant associated with the Clovis culture who was found in Montana very close to the present-day Canadian border. The Clovis infant came from a lineage different from that of present-day inhabitants of neighboring Canada, reflecting major population movements that must have happened later.

In some places in the Americas, ancient DNA confirms the theory that populations have remained in the same region for thousands of years. According to analyses we and Lars Fehren-Schmitz have done of Peruvians dating up to nine thousand years ago, there has been broad continuity in Native American populations in this region. All the ancient genomes from Peru that we have studied are more closely related to each other and to present-day Native Americans from Peru who speak the Quechua and Aymara languages than they are to any other present-day South American populations. We have similar findings from Native American individuals from southern Argentina dating to around eight thousand years ago, and Native American individuals from southern Brazil dating to around ten thousand years ago. The same applies to Native Americans from the islands off British Columbia, who appear to have been part of a continuous population for around six thousand years, even if the local continuity does not clearly go back more than ten thousand years.[31] All are more closely related to Native Americans who live in the same regions today than to Native Americans far away.

The Genomic Rehabilitation of Joseph Greenberg

The genetic discovery of the spread of the First Americans also helps to resolve a linguistic controversy. The extraordinary diversity of Native American languages had been noted as early as the seventeenth century, with some European missionaries attributing it to the devil's efforts to resist the conversion of Native populations by making the language that missionaries needed to learn to proselytize to one population useless for proselytizing to the next. Linguists can be divided into "splitters," who emphasize differences among languages, and "lumpers," who emphasize their common roots. One of the most extreme splitters was Lyle Campbell, who divided about one thousand Native American languages into about two hundred families (groups of related languages), sometimes even localized to particular river valleys.[32] One of the most extreme lumpers was Joseph Greenberg, who argued that he could group all Native American languages into just three families, the deep connections of which he could trace. He argued that these three families reflected three great waves of migration from Asia.

Campbell and Greenberg clashed famously in their interpretation of Native American language relationships, with Campbell finding Greenberg's tripartite classification so objectionable that he wrote in 1986 that Greenberg's classification "should be shouted down."[33] In fact, two of the language families are indisputable: Eskimo-Aleut languages spoken by many of the indigenous peoples of Siberia, Alaska, northern Canada, and Greenland, and Na-Dene languages spoken by a subset of the Native American tribes living on the Pacific coast of northern North America, in the interior of northern Canada, and in the southwestern United States.

But it was Greenberg's third family, "Amerind," which he claimed includes about 90 percent of the languages of Native Americans, that so many linguists found objectionable. The method that Greenberg used to propose Amerind was to study several hundred words across different Native American languages and to score them according to the extent to which they were shared. By finding high rates of sharing, he claimed evidence for common origin. As he saw it, proto-Amerind was spoken by the first Americans south of the ice sheets. Because he found that every non-Na-Dene and non-Eskimo-Aleut language throughout the Americas could be classified as Amerind using this approach, he concluded that the language data supported a theory of three great waves of Native American dispersal from Asia. If there had been another wave, it would have left another distinct set of languages.

The critique of Greenberg's ideas that followed was withering. Critics argued that the list of words was too brief to establish commonality. Critics also disputed the claim that these words truly stemmed from common roots. Identification of shared words is thought to become difficult for time depths of more than a few thousand years because languages change so fast, but Greenberg was claiming to detect links at twice this time depth.

But Greenberg got something right. His category of Amerind corresponds almost exactly to the First American category found by genetics. The clusters of populations that he predicted to be most closely related based on language were in fact verified by the genetic patterns in populations for which data are available. And the present-day balkanization of Native American languages also reflects a history in which the great majority of populations descend from a single

All Native American groups today have large proportions of First American ancestry.

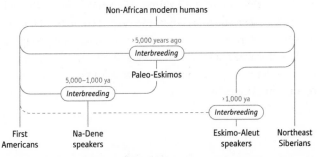

Figure 20. This simplified tree relates the three groupings of Native American populations hypothesized by Joseph Greenberg based on linguistic data. The groupings correspond to three distinct entries into the Americas, but Greenberg did not know about the high proportions of First American ancestry in all groups: about 90 percent in Na-Dene speakers and about 60 percent in Eskimo-Aleut speakers.

migratory spread. Anyone looking at a language map of the Americas can see that its appearance is qualitatively different from that of Eurasia or Africa, with dozens of language families restricted to small territories, compared to the vast swaths of territory in Eurasia and Africa inhabited by people who speak closely related tongues in the Indo-European, Austronesian, Sino-Tibetan, and Bantu language families, each of which reflects a history of mass migrations and population replacements. The First American expansion seems to have been so fast that the languages of the continent are related by a rake-like structure with many tines extending in parallel to a common root that dates close to the time of the early settlement of the Americas.[34] So both the genetic and linguistic evidence support a scenario in which many of the present-day Native American populations are direct descendants of populations that plausibly lived in the same region shortly after the first peopling of the continent. This suggests that after the initial dispersal, population replacement was more infrequent in the Americas than it was in Africa and Eurasia.

While the genetic data provided a large measure of confirmation for Greenberg's broad picture, he missed something important. Although Eskimo-Aleut and Na-Dene speakers are genetically distinguishable from other Native Americans because they carry ancestry from distinct streams of migration from Asia, both have large amounts of First American ancestry: around a 60 percent mixture proportion in the case of the Eskimo-Aleut speakers we studied, and around a 90 percent proportion in the case of some Na-Dene speakers.[35] So while Greenberg's three predicted language groups correlate well with three ancient populations, First Americans have made a dominant demographic contribution to all present-day indigenous peoples in the Americas.

Population Y

The next card dealt from the genetic deck was a complete surprise—at least to us geneticists.

Some physical anthropologists studying the shapes of human skeletons had for years been asserting that there are some American skeletons, dating to before ten thousand years ago, that do not look like what one would expect for the ancestors of today's Native Americans. The most iconic is Luzia, an approximately 11,500-year-old skeleton whose remains were found in Lapa Vermelha, Brazil, in 1975. Many anthropologists find the shape of her face more similar to those of indigenous peoples from Australia and New Guinea than to those of ancient or modern peoples of East Asia, or Native Americans. This puzzle led to speculation that Luzia came from a group that preceded Native Americans. Anthropologist Walter Neves has identified dozens of Mesoamerican and South American skeletons with what he calls a "Paleoamerican" morphology. Exhibit number one for Neves is a set of fifty-five skulls dating to ten thousand years ago or more from a prehistoric garbage dump at Lagoa Santa in Brazil.[36]

These claims are controversial. Morphological traits vary depending on diet and environment, and after the arrival of humans in the Americas, natural selection as well as random changes that accumulate in populations over time may have contributed to morphological change. The experience of Kennewick Man, whose skeleton has morphological affinities to those of Pacific Rim populations but

A Deep Connection Between Amazonia and Australasia

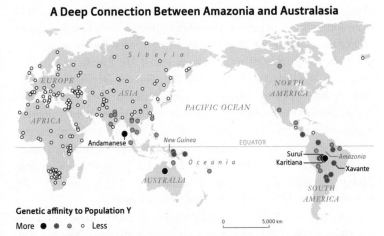

Genetic affinity to Population Y

More ● ● ● ○ Less   0 _____ 5,000 km

Figure 21. Despite extraordinary geographic distance, populations in the Amazon share ancestry with Australians, New Guineans, and Andamanese to a greater extent than with other Eurasians. This may reflect an early movement of humans into the Americas from a source population that is no longer substantially represented in northeast Asia.

genetically is derived entirely from the same ancestral population as other Native Americans, serves as a great warning—an object lesson about the danger of interpreting morphology as strong evidence of population relationships.[37] Many have criticized Neves by suggesting that his analyses were statistically flawed, in that he chose which sites to include in his analysis in order to strengthen his Paleoamerican idea and deliberately left out those that did not fit, an approach inconsistent with rigorous science.

Nonetheless, Pontus Skoglund decided to inspect Native American genetic data more closely, looking for traces of ancestry different from the First Americans. His logic went as follows. If there were ancient people on the continent who were displaced by First Americans, they may have mixed with the ancestors of present-day populations, leaving some statistical signal in the genomes of people living today.

Skoglund undertook a Four Population Test to compare all possible pairs of populations from the Americas that we had previously thought were entirely of First American ancestry to all possible pairs of populations outside the Americas, among them indigenous people from Australasia (including Andaman Islanders, New Guineans, and Australians) and other populations hypothesized by some anthropologists to be related to Paleoamericans. He found two Native American populations, both from the Amazon region of Brazil, that are more closely related to Australasians than to other world populations. After joining my laboratory as a postdoctoral scientist, Skoglund found weaker signals of genetic affinity to Australasians, but still probably real, in other Native American populations ringing the Amazon basin. He estimated that the proportion of ancient ancestry in these populations was small—1 to 6 percent—with the rest being consistent with First American ancestry.[38]

Skoglund and I were initially skeptical about these findings, but the statistical evidence just kept getting stronger. We saw the same patterns in multiple independently collected datasets. We also showed that these patterns could not arise as a result of recent migrations from Asian populations—while Amazonians had their strongest affinity to indigenous people from Australia, New Guinea, and the Andaman Islands (compared to East Asians as a baseline), they were not particularly close to any of them. Also contradicted by the genetic data was a Polynesian migration from the Pacific across to the Americas. While such a migration could have reasonably occurred over the past couple of thousand years as Polynesians mastered the technology of transoceanic travel, the affinities we found had nothing in common with Polynesians. It really looked like evidence of a migration into the Americas of an ancient population more closely related to Australians, New Guineans, and Andamanese than to present-day Siberians. We concluded that we had found evidence of a "ghost" population: a population that no longer exists in unmixed form. We called this "Population Y" after the word *ypykuéra*, meaning "ancestor" in Tupí, the language family of the populations with the largest proportions of this ancestry.

The Tupí-speaking population in which we found the most Population Y ancestry was the Suruí, the authors of the origin myth that begins this chapter. They now number about fourteen hundred people and live in the Brazilian state of Rondônia.[39] They have been relatively isolated, establishing formal relations with the government of Brazil only in the 1960s when road builders came through their territory. Since then, the Suruí have defended their land from deforestation, taken over coffee plantations, and reported illegal loggers and miners. They have sought representation from indigenous rights groups in the United States and claimed carbon credits for the greenhouse gases conserved through the rainforest they have protected.

Another group belonging to the Tupí language family in which we found Population Y ancestry is the Karitiana. The Karitiana are discussed at the beginning of this chapter as one of the first Native American tribes to become active in protesting against genetic research—in their case because of concern that DNA samples had been taken from them in 1996 with a promise of improved access to health care that has never been realized. The Karitiana are around three hundred strong and also come from Rondônia. The samples we analyzed were not part of this tainted 1996 sampling but instead from a 1987 sampling in which informed consent procedures consistent with the ethical standards of the time appear to have been followed. I hope that the Karitiana individuals who encounter our findings will welcome these observations about their distinctive ancestry as a positive discovery that highlights benefits that can come from engaging in scientific studies.[40]

The third population in which we found substantial Population Y ancestry is the Xavante, who speak a language of the Ge group, which is different from the Tupí language group spoken by the Suruí and Karitiana. They number around eighteen thousand people and are located in Brazil's Mato Grosso state, on the Brazilian plateau. They have been forcibly relocated, their land today suffers from environmental degradation, and their indigenous way of life is constantly under threat from development.[41]

We found little or no Population Y ancestry in Mesoamerica or in South Americans to the west of the high Andes. We also did not detect Population Y ancestry in the almost thirteen-thousand-year-old genome of the Clovis culture infant from the northern United States, or in present-day Algonquin speakers from Canada. The Population Y geographic distribution is largely limited to Amazonia, providing yet more evidence for an ancient origin. The fact that

Population Y ancestry is restricted to difficult terrain far from the Bering link to Asia is perhaps what one would expect from an original pioneering population that was once more broadly distributed and was then marginalized by the expansion of other groups. This pattern mirrors the distribution of some other language families—for example, the Tuu, Kx'a, and Khoe-Kwadi languages spoken by the Khoe and San in southern Africa—where islands of these speakers in rugged terrain are surrounded by seas of people speaking other languages.

The fact that the strongest statistical evidence of the ancient lineage we detect is in Brazil, the home of "Luzia" and the Lagoa Santa skeletons, is remarkable, but does not prove that the ancient lineage we discovered coincides with the "Paleoamerican" morphology hypothesized by Neves and others. Neves claimed to see the Paleoamerican morphology not only in ancient Brazilians but also in ancient and relatively recent Mexicans, and yet we found no hint of a signal in Mexicans. In addition, Eske Willerslev's group obtained DNA from two Native American groups that had skeletal morphology typical of Paleoamericans according to Neves: Pericúes in the Baja California peninsula of northwestern Mexico and Fuegans in the southern tip of South America. Neither of these groups carried Population Y ancestry.[42]

What, then, does the genetic pattern mean? We already know from archaeology that humans probably arrived south of the ice sheets before the opening of the ice-free corridor, leaving remains at archaeological sites including Monte Verde and the Paisley Caves. But the big population explosion, marked by the Clovis people, only occurred once the ice-free corridor had opened. The genetic data could be giving evidence of early peopling of the Americas by a minimum of two very different groups moving in from Asia, perhaps along two different routes and at different times. If Population Y spread through parts of South America before the First Americans, then it seems likely that after this initial peopling, the First Americans advanced into nearly all of the territories the Population Y people had already visited, replacing them either completely or only partially, as in Amazonia. Population Y ancestry may have survived better in Amazonia than it did elsewhere because of the relative impenetrability of the Amazonian environment. This could have slowed down the movement of First Americans into the region enough to allow people living there to mix with the new migrants rather than simply being replaced.

The Australasian-related ancestry in the Suruí today amounts to a small percentage—about the same as the Neanderthal ancestry in all non-Africans—but it would be unwise to dismiss its importance. This is because the impact of Population Y on Amazonians may be much greater than 2 percent. The ancestors of Population Y had to traverse enormous spaces in Siberia and northern North America where the ancestors of First Americans were also living. It is likely that Population Y was already mixed with large amounts of First American–related ancestry when it started expanding into South America. If so, then the ancestry derived from a lineage related to southern Asians is only a kind of "tracer dye" for Population Y ancestry—like the heavy metals injected into patients' veins in hospitals to track the paths of their blood vessels in an MRI scan. Our estimate of around 2 percent Population Y ancestry in the Suruí is based on the assumption that Population Y traversed the entirety of Northeast Asia and America without mixing with other people it encountered. If we allow for the likelihood that there was mixture with populations related to First Americans on the way, the proportion of Population Y in the Suruí could be as high as 85 percent and still produce the observed statistical evidence of relatedness to Australasians. If the true proportion is even a fraction of this, then the story of First Americans expanding into virgin territory is profoundly misleading. Instead, we need to think in terms of an expansion of a highly substructured founding population of the Americas. The history and timing of the arrival of Population Y in the Americas is likely to be resolved only with recovery of ancient DNA from skeletons with Population Y ancestry.

After the First Americans

The great promise of genetic data lies not only in what they can tell us about the deepest origins of Native Americans but also in what genetic data has to say about more recent times and how populations got to be the way they are today.

A prime example is insight into the origin of speakers of Na-Dene languages, who live along the Pacific coast of North America, in parts of northern Canada, and as far south as Arizona in the United States. The overwhelming consensus among linguists is that these languages stem from an ancestral language no more than a few thousand years old, and that their dispersal over this vast range in northwestern America must have been driven at least in part by migrations. In an astonishing development in 2008, the American linguist Edward Vajda documented a deeper connection between Na-Dene languages and a language family of central Siberia called Yeniseian, once spoken by many populations, though today only the Ket language of the Yeniseian family is still used on a day-to-day basis.[43] These results suggest that despite the enormous distance, a relatively recent migration from Asia gave rise to Na-Dene speakers in the Americas.

What new information does genetics add? Our 2012 study found that the Na-Dene-speaking Chipewyan carry a type of ancestry not shared with many other Native Americans, providing evidence for the later Asian migration theory.[44] We estimated that this ancestry constituted only around 10 percent of Chipewyan ancestry, but it was striking all the same. We wondered whether we could use this distinctive strain of ancestry in the Chipewyans as a tracer dye to document an ancestral link between Na-Dene speakers like Chipewyans and individuals from past archaeological cultures who could be studied with ancient DNA.

In 2010, Eske Willerslev and colleagues published genome-wide data from an approximately four-thousand-year-old lock of hair taken from a frozen individual of the Saqqaq culture, the first human culture of Greenland.[45] Their analysis showed that this man belonged to a population that had a distinct blend of ancestry compared both to First Americans in the south and the Eskimo-Aleuts who followed them in the Arctic. Willerslev's group expanded its claim in 2014 when it reported data from several additional "Paleo-Eskimos," as people who preceded Eskimo-Aleuts are called by archaeologists.[46] All these individuals were broadly related, and the authors argued that they represented a distinct migration from Asia that was different from all prior and subsequent ones. They argued that the Paleo-

Eskimos largely went extinct without leaving descendants after the arrival of Eskimo-Aleut speakers around fifteen hundred years ago.

In our 2012 study, we tested the idea that the Paleo-Eskimos exemplified by the Saqqaq individual were descended from a distinct migration to the Americas. To our surprise, we found no statistical evidence for a distinct migration. Instead, our tests were consistent with the possibility that the Saqqaq derived their ancestry from the same source that contributed to the Na-Dene-speaking Chipewyans, just in different proportions. Since we know from genetic data that only around 10 percent of the ancestry of many Na-Dene speakers today is from this late Asian migration, it is easy to understand why the clustering analysis used by Willerslev's team missed the connection to Na-Dene speakers. We proposed that the Na-Dene and Saqqaq might both derive part of their ancestry from the same ancient migration from Asia to the Americas.

In 2017, Pavel Flegontov, Stephan Schiffels, and I confirmed that the Paleo-Eskimo lineage did not die out, and instead lives on in the Na-Dene.[47] By examining rare mutations that reflect recent sharing between diverse Native American and Siberian populations, we found evidence for recent common ancestors between the ancient Saqqaq individual and present-day Na-Dene. In fact, the hypothesis that Paleo-Eskimo lineages went extinct after the arrival of Eskimo-Aleut speakers is even more profoundly wrong than I had originally suggested in my 2012 paper.[48] The correct way to view the ancestry of present-day speakers of Eskimo-Aleut languages is as a mixture of lineages related to Paleo-Eskimos and First Americans. In other words, far from being extinct, the population that included Paleo-Eskimos lives on in mixed form not just in Na-Dene speakers, but also in Eskimo-Aleut speakers.

Our 2017 work also revealed an entirely new and unifying way to view the deep ancestry of the peoples of the Americas. In this new vision, there were just two ancestral lineages that contributed all Native American ancestry apart from that in Population Y: the First Americans and the population that brought new small stone tools and the first archery equipment to the Americas around five thousand years ago and founded the Paleo-Eskimos.[49] We could show this because, mathematically, we can fit a model to the data in which all Native Americans excluding Amazonians with their Population Y ancestry can be described as mixtures of two ancestral populations related differentially to Asians. Mixtures of these two ancestral populations produced the three source populations that migrated from Asia to America and that are associated with Eskimo-Aleut languages, Na-Dene, and all other languages.

A second genetic revelation about Native American population history is clearest in the Chukchi, a population of far northeastern Siberia that speaks a language unrelated to any spoken in the Americas. My analyses revealed that the Chukchi harbor around 40 percent First American ancestry due to backflow from America to Asia.[50] For those who are dubious about the idea that descendants of First Americans could have reexpanded out of America and then made a substantial demographic impact on Asia—who are used to thinking about the migratory path between Asia and America as a one-way street—it might be tempting to argue that the genetic affinity of the Chukchi to Native Americans simply reflects that they are the closest cousins of the First Americans in Asia. This bias also impeded my own thinking for more than a year as I tried to make sense of the data we had from diverse Native Americans. But the genetic data clarify that the affinity is due to back-migration, as the Chukchi are more closely related to some populations of entirely First American ancestry than to others, a finding that can only be explained if a sublineage of First Americans that originated well after the initial diversification of First American lineages in North America migrated back to Asia. The explanation for this observation is that the Eskimo-Aleut speakers who established themselves in North America mixed heavily with local Native Americans (who contributed about half their ancestry) and then took their successful way of life back through the Arctic with them to Siberia, contributing not only to the Chukchi but also to local speakers of Eskimo-Aleut languages. The identification of a reflux of First American ancestry into Asia—a type of finding that is difficult to prove with archaeology—is the kind of surprise that genetics is in a unique position to deliver.

A third example of what genetics can offer is the story of the arrival of agriculture to the U.S. Southwest from northern Mexico. Today, these regions are linked by a widespread language family called Uto-

Aztecan, which linguists have traditionally viewed as having spread from north to south, based on the fact that most of the languages in this group and some plant names that are shared across the languages are typical of the northern end of the present-day Uto-Aztecan distribution. However, others have argued that the languages radiated northward from Mexico, following the spread of maize agriculture. It has been suggested, most forcefully by the archaeologist Peter Bellwood, that languages and peoples tend to move with the spread of agriculture.[51] Studying the ancient DNA of people who lived before and after the arrival of maize in the region, along with comparison to the present-day inhabitants, can test this theory at least in part. We are beginning to find some clues in ancient DNA. Studies of ancient maize have now shown that this crop first entered the U.S. Southwest by a highland route (inland, over hills) more than four thousand years ago, and then was replaced by strains of maize of a lowland coastal origin around two thousand years ago.[52] This is a remarkable example of how plants, too, have had histories of migration and recurrent mixture, although in the case of domesticated crops the migrations and mixtures are if anything likely to be more dramatic because humans have subjected crops to artificial selection. It will only be a matter of time before we are able to test whether new peoples moved with the new crops.

The dream, of course, is to carry out studies like these more systematically. Modern genetic studies and ancient DNA enable us to discover how Native American cultures are connected by links of migration, and how the spread of languages and technologies corresponded to ancient population movements. Many of these stories have been lost because of the European exploitation that has decimated Native American populations and their culture. Genetics offers the opportunity to rediscover lost stories, and has the potential to promote not just understanding but also healing.

East Asia and the Pacific

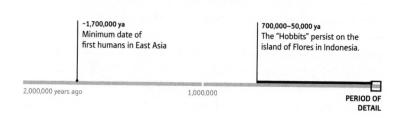

~1,700,000 ya
Minimum date of
first humans in East Asia

700,000–50,000 ya
The "Hobbits" persist on the
island of Flores in Indonesia.

2,000,000 years ago

1,000,000

**PERIOD OF
DETAIL**

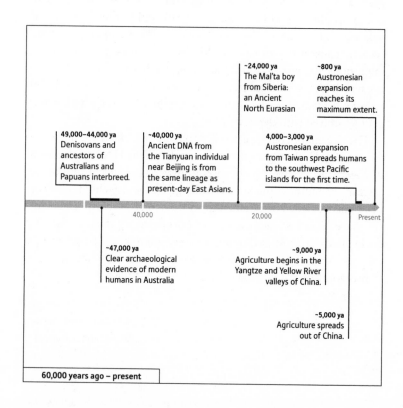

~24,000 ya
The Mal'ta boy
from Siberia:
an Ancient
North Eurasian

~800 ya
Austronesian
expansion
reaches its
maximum extent.

49,000–44,000 ya
Denisovans and
ancestors of
Australians and
Papuans interbreed.

~40,000 ya
Ancient DNA from
the Tianyuan individual
near Beijing is from
the same lineage as
present-day East Asians.

4,000–3,000 ya
Austronesian expansion
from Taiwan spreads humans
to the southwest Pacific
islands for the first time.

40,000

20,000

Present

~47,000 ya
Clear archaeological
evidence of modern
humans in Australia

~9,000 ya
Agriculture begins in the
Yangtze and Yellow River
valleys of China.

~5,000 ya
Agriculture spreads
out of China.

60,000 years ago – present

The Genomic Origins of East Asians

The Failure of the Southern Route

East Asia—the vast region encompassing China, Japan, and Southeast Asia—is one of the great theaters of human evolution. It harbors more than one third of the world's population and a similar fraction of its language diversity. Pottery was first invented there at least nineteen thousand years ago.[1] It was the jumping-off point for the peopling of the Americas before fifteen thousand years ago. East Asia witnessed an independent and early invention of agriculture around nine thousand years ago.

East Asia has been home to the human family for at least around 1.7 million years, the date of the oldest known *Homo erectus* skeleton found in China.[2] The earliest human remains excavated in Indonesia are similarly old.[3] Archaic humans—whose skeletal form is not the same as that of humans whose anatomically modern features begin to appear in the African fossil record after around three hundred thousand years ago[4]—have lived in East Asia continuously since those times. For example, genetic evidence shows that the Denisovans mixed with ancestors of present-day Australians and New Guineans shortly after fifty thousand years ago. And archaeological and skeletal evidence shows that the one-meter-tall "Hobbits" also persisted until around this same time on Flores island in Indonesia.[5]

There has been intense debate about the extent to which the archaic humans of East Asia contributed genetically to people living today. Chinese and Western geneticists nearly all agree that present-day humans outside of Africa descend from a dispersal after around fifty thousand years ago, which largely displaced previously established human groups.[6] Some Chinese anthropologists and archaeologists, on the other hand, have documented similarities in skeletal features and stone tool styles in people who lived in East Asia before and after this time, raising the question of whether there has been some degree of continuity.[7] At the time of this writing, our knowledge of East Asian population history is relatively limited compared to that of West Eurasia because less than 5 percent of published ancient DNA data comes from East Asia. The difference reflects the fact that ancient DNA technology was invented in Europe, and it is nearly impossible for researchers to export samples from China and Japan because of government restrictions or a preference that studies be led by local scientists. This has meant that these regions have missed out on the first few years of the ancient DNA revolution.

In the west, the grand narrative is that sometime after around fifty thousand years ago, modern humans began making sophisticated Upper Paleolithic stone tools, which are characterized by narrow stone blades struck in a new way from pre-prepared cores. The Near East is the earliest known site of Upper Paleolithic stone tools, and this technology spread rapidly to Europe and northern Eurasia. It would be natural to expect, given how successful the people who made Upper Paleolithic technology were, that this know-how would have overspread East Asia too. But that is not what happened.

The archaeological pattern in the east does not conform to that in the west. Around forty thousand years ago and across a vast tract of land in China and east of India there is indeed archaeological evidence of great behavioral change associated with the arrival of modern humans, including the use of sophisticated bone tools, shell beads or perforated teeth for body decoration, and the world's earliest known cave art.[8] In Australia, archaeological evidence of human campsites makes it clear that modern humans arrived there at least by about forty-seven thousand years ago,[9] which is about as old as the earliest evidence for modern humans in Europe.[10] So it is absolutely clear that modern humans arrived in East Asia and Australia around the same time as they came to Europe. But, puzzlingly, the first modern humans in central and southern East Asia, and those in Australia, did not use Upper Paleolithic stone tools. Instead, they used other technologies, some of which were more similar to those used by modern humans in Africa tens of thousands of years earlier.[11]

Prompted by these observations, the archaeologists Marta Mirazon Lahr and Robert Foley argued that the first humans in Australia might derive from a migration of modern humans out of Africa and the Near East prior to the development of Upper Paleolithic technology in the west. According to this "Southern Route" hypothesis, the migrants left Africa well before fifty thousand years ago and skirted along the coast of the Indian Ocean, leaving descendants today among the indigenous people of Australia, New Guinea, the Philippines, Malaysia, and the Andaman Islands.[12] The anthropologist Katerina Harvati and colleagues also documented skeletal similarities between Australian Aborigines and Africans that, they argued, provide evidence for this model.[13]

The Southern Route hypothesis was far more than a claim that there were modern humans outside of Africa well before fifty thousand years ago—a fact that every serious scholar now accepts.[14] Evidence of early modern humans outside of Africa well before fifty thousand years ago includes the morphologically modern skeletons in Skhul and Qafzeh in present-day Israel that date to between around 130,000 to 100,000 years ago.[15] Stone tools found at the site of Jebel Faya from around 130,000 years ago are similar to ones found in northeast Africa from around the same time, suggesting that modern humans made an early crossing of the Red Sea into Arabia.[16] There is also tentative genetic evidence of an early impact of modern humans outside Africa, with Neanderthal genomes harboring a couple of percent of ancestry that may derive from interbreeding with a modern human lineage that separated a couple of hundred thousand years ago from present-day human lineages, as expected if a modern human population possibly related to that in Skhul and Qafzeh interbred with Neanderthal ancestors.[17] Although many geneticists, including me, are still on the fence about whether this finding of earlier interbreeding between modern humans and Neanderthals is compelling, the key point is that almost all scholars now agree that there were early dispersals of modern humans into

Asia that preceded the widely accepted dispersals after fifty thousand years ago that contributed in a major way to all present-day non-Africans. The outstanding question raised by the Southern Route hypothesis is not whether such expansions occurred, but whether they had an important long-term impact on humans living today.

In 2011, Eske Willerslev led a study that seemed to show that the early expansions indeed left an impact.[18] He and his colleagues reported a Four Population Test showing that Europeans share more mutations with East Asians than with Aboriginal Australians, as would be expected from a Southern Route contribution to the lineage of Australians. Applying a Southern Route migration model to the genomic data, they estimated that Australian Aborigines harbor ancestry from a modern human population that split from present-day Europeans at twice the time depth that East Asian ancestors split from Europeans (seventy-five thousand to sixty-two thousand years ago versus thirty-eight thousand to twenty-five thousand years ago).

There was a problem, though, which is that the analysis did not account for the 3 to 6 percent of ancestry that Australians inherited from archaic Denisovans.[19] Because Denisovans were so divergent from modern humans, mixture from them could cause Europeans to share more mutations with Chinese than with Australian Aborigines. Indeed, this explained the findings. My laboratory showed that after accounting for Denisovan mixture, Europeans do not share more mutations with Chinese than with Australians, and so Chinese and Australians derive almost all their ancestry from a homogeneous population whose ancestors separated earlier from the ancestors of Europeans.[20] This revealed that a series of major population splits in the history of non-Africans occurred in an exceptionally short time span—beginning with the separation of the lineages leading to West Eurasians and East Eurasians, and ending with the split of the ancestors of Australian Aborigines from the ancestors of many mainland East Eurasians. These population splits all occurred after the time when Neanderthals interbred with the ancestors of non-Africans fifty-four to forty-nine thousand years ago, and before the time when Denisovans and the ancestors of Australians mixed, genetically estimated to be 12 percent more recent than the Neanderthal/modern human admixture, that is, forty-nine to forty-four thousand years ago.[21]

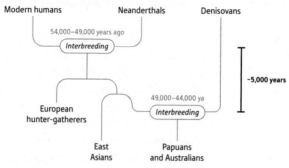

Two Major Population Splits Within About 5,000 Years

Figure 22. Two major splits were sandwiched in an approximately five-thousand-year period between the Neanderthal and Denisovan interbreeding events.

The rapid succession of lineage separations during the relatively short interval between Neanderthal and Denisovan interbreeding with modern humans suggests that throughout Eurasia, modern humans were moving into new environments where their technology or lifestyle allowed them to expand, displacing the previously resident groups. The spread was so fast that it is hard to imagine that archaic humans who had already been resident there for close to two million years, and who we know were also there when modern humans expanded based on the evidence of interbreeding with Denisovans, put up much resistance. Even if early modern humans expanded into East Asia via a Southern Route, they were likely also replaced by later waves of human migrants and can be ruled out as having contributed more than a very small percentage of the ancestry of present-day people.[22] In East Asia as in West Eurasia, the expansion of modern humans out of Africa and the Near East had an effect akin to the erasing of a blackboard, creating a blank slate for the new people. The old populations of Eurasia collapsed, and in their place came new groups that swiftly inhabited the landscape. There is no genetic evidence of any substantial ancestry from these earlier populations in present East Asians.[23]

So if essentially all modern human ancestry in East Asia and Australia today derives from the same group that contributed to West

Eurasians, what explains how Southeast Asians and Australians missed out on the Upper Paleolithic technology that is so tightly linked with the spread of modern human populations into the Near East and Europe?

The first long-bladed stone tools characteristic of Upper Paleolithic technology in the archaeological record date to between fifty thousand and forty-six thousand years ago.[24] But genetically, the split of the lineages leading to West Eurasians and East Asians may have been more ancient since, as I have discussed, it almost certainly occurred within a few thousand years after the admixture of modern humans with Neanderthals fifty-four thousand to forty-nine thousand years ago. So the main split of West Eurasian and East Asian ancestors could have occurred before the development of Upper Paleolithic technology, and the geographic distribution of this technology could just reflect the spread of the population that invented it.

There is a piece of corroborating evidence for the theory that Upper Paleolithic technology developed after the split of the main lineages leading to West Eurasians and East Asians. The Ancient North Eurasians, known earliest from the approximately twenty-four-thousand-year-old remains of the boy from the Mal'ta site in eastern Siberia,[25] are on the lineage leading to West Eurasians, which has always been puzzling for geneticists because the Ancient North Eurasians lived geographically closer to East Asia. But it makes sense in light of the geographic distribution of Upper Paleolithic stone tools, which are associated not just with West Eurasians but also North Eurasians and Northeast Asians. Both the distributions of stone tool technology and of genetic ancestry are as expected if Upper Paleolithic technology came into full flower in a population that lived prior to the separation of the lineages leading to Ancient North Eurasians and West Eurasians, but after the separation of the lineage leading to East Asians.

Whatever the reason for the fact that Upper Paleolithic technology never spread to southern East Asia, it is clear from what happened next, and the success these people had in displacing the previously resident populations such as Denisovans, that Upper Paleolithic technology itself was not essential to the successful spread of modern humans into Eurasia after around fifty thousand years ago. It was something more profound than Upper Paleolithic stone tool technology—an inventiveness and adaptability of which the technology was just a manifestation—that allowed these expanding modern humans to prevail everywhere, including in the east.

The Beginnings of Modern East Asia

The first genomic survey of modern East Asian populations was published in 2009, and reported data on nearly two thousand individuals from almost seventy-five populations.[26] The authors focused on their finding that human diversity is greater in Southeast Asia than in Northeast Asia. They interpreted this pattern as evidence of a single wave of modern humans reaching Southeast Asia and then spreading from there northward into China and beyond, following a model in which the genetic diversity of present-day populations can be accounted for by a single population moving out of Africa and spreading in all directions, losing genetic diversity as each small pioneer group budded off.[27] But we now know that this model is likely to be of limited use. In Europe there have been multiple population replacements and deep mixtures, and we now know from ancient DNA that present-day patterns of diversity in West Eurasia provide a distorted picture of the first modern human migrations into the region.[28] The model of a south-to-north migration, losing diversity along the way, is profoundly wrong for East Asia.

In 2015 Chuanchao Wang arrived in my laboratory bearing a treasure: genome-wide data from about four hundred present-day individuals from about forty diverse Chinese populations. China had been sparsely sampled in DNA studies because of regulations limiting the export of biological material. Wang and his colleagues therefore did the genetic work in China, and he brought the data to us electronically. Over the next year and a half, we analyzed these data together with more data from other East Asian countries that had previously been published and with ancient DNA from the Russian Far East generated in our lab. This allowed us to come up with new genetic insights about the deep population history of East Asia and the origins of its current inhabitants.[29]

By using a principal component analysis, we found that the ancestry of the great majority of East Asians living today can be described by three clusters.

The first cluster is centered on people currently living in the Amur River basin on the boundary between northeastern China and Russia. It includes ancient DNA data that my laboratory and others had obtained from the Amur River basin. So, this region has been inhabited by genetically similar populations for more than eight thousand years.[30]

The second cluster is located on the Tibetan Plateau, a vast area north of the Himalayas, much of which is at a higher altitude than the tallest of the European Alps.

The third cluster is centered in Southeast Asia, and is most strongly represented by individuals from indigenous populations living on the islands of Hainan and Taiwan off the coast of mainland China.

We used Four Population Test statistics to evaluate models of the possible relationships among present-day populations representing these clusters and Native Americans, Andaman islanders, and New Guineans. The latter three populations have been largely isolated from the ancestors of mainland East Asians at least since the last ice age, and their East Asian–related ancestry effectively serves as ancient DNA from that period.

Our analysis supported a model of population history in which the modern human ancestry of the great majority of mainland East Asians living today derives largely from mixtures—in different proportions—of two lineages that separated very anciently. Members of these two lineages spread in all directions, and their mixture with each other and with some of the populations they encountered transformed the human landscape of East Asia.

The Ghost Populations of the Yangtze and Yellow Rivers

One of the handful of places in the world where farming independently began was China. Archaeological evidence shows that starting around nine thousand years ago, farmers started tilling the wind-

blown sediments near the Yellow River in northern China, growing millet and other crops. Around the same time, in the south near
the Yangtze River, a different group of farmers began growing other
crops, including rice.[31] Yangtze River agriculture expanded along two
routes—a land route that reached Vietnam and Thailand beginning
around five thousand years ago, and a maritime route that reached the
island of Taiwan around the same time. In India and in central Asia,
Chinese agriculture collided for the first time with the expansion of
agriculture from the Near East. Language patterns also hint at the
possibility of movements of people. Today the languages of mainland East Asia comprise at least eleven major families: Sino-Tibetan,
Tai-Kadai, Austronesian, Austroasiatic, Hmong-Mien, Japonic, Indo-
European, Mongolic, Turkic, Tungusic, and Koreanic. Peter Bellwood has argued that the first six correspond to expansions of East
Asian agriculturalists disseminating their languages as they moved.[32]

What can we say based on the genetics? Because of restrictions on
exporting skeletal material from China, the information that genetic
data currently provide about the deep population history of East Asia
is far behind that of West Eurasia or even of America. Nevertheless,
Wang learned what he could based on the little ancient DNA data
we had and patterns of variation in present-day people.

We found that in Southeast Asia and Taiwan, there are many
populations that derive most or all of their ancestry from a homogeneous ancestral population. Since the locations of these populations strongly overlap with the regions where rice farming expanded
from the Yangtze River valley, it is tempting to hypothesize that
they descend from the people who developed rice agriculture. We
do not yet have ancient DNA from the first farmers of the Yangtze
River valley, but my guess is that they will match this reconstructed
"Yangtze River Ghost Population," the name that we have given the
population that contributed the overwhelming majority of ancestry
to present-day Southeast Asians.

But we found that the Han Chinese—the world's largest group
with a census size of more than 1.2 billion—is not consistent with
descending directly from the Yangtze River Ghost population.
Instead, the Han also have a large proportion of ancestry from another
deeply divergent East Asian lineage. The highest proportions of this

other ancestry are found in northern Han, consistent with work since 2009 that has shown that the Han harbor subtle differences along a north-to-south gradient.[33] This pattern is as expected from a history in which the ancestors of the Han radiated out from the north and mixed with locals as they spread south.[34]

What could the other ancestry type be? The Han, who unified China in 202 BCE, are believed based on historical sources to have emerged from the earlier Huaxia tribes who themselves emanated from earlier groups in the Yellow River Valley of northern China. This was one of the two Chinese regions where farming originated, and it is also the place from which farming spread to the eastern Tibetan Plateau beginning around thirty-six hundred years ago.[35] Since the Han and Tibetans are also linked by their Sino-Tibetan languages, we wondered whether they might share a distinctive type of ancestry as well.

When Wang built his model of deep East Asian population history, he found that the Han and Tibetans both harbored large proportions of their ancestry from a population that no longer exists in unmixed form and that we could exclude as having contributed ancestry to many Southeast Asian populations. Because of the combined evidence of archaeology, language, and genetics, we called this the "Yellow River Ghost Population," hypothesizing that it developed agriculture in the north while spreading Sino-Tibetan languages. Ancient DNA from the first farmers of the Yellow River Valley will reveal whether this conjecture is correct. Once available, ancient DNA will also make it possible to learn about features of East Asian population history that are impossible to discern based on analysis only of populations living today, whose deep history has been clouded by many additional layers of migration and mixture.

The Great Admixtures at the East Asian Periphery

Once the core agricultural populations of the Chinese plain—the Yangtze and Yellow River ghost populations—formed, they expanded in all directions, mixing with groups that had arrived in earlier millennia.

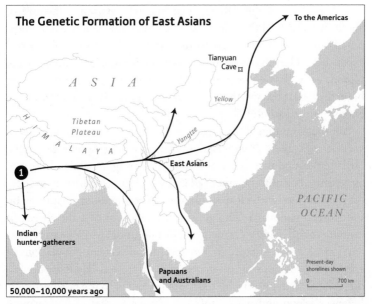

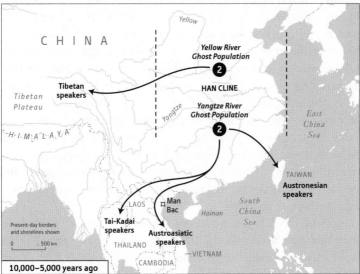

Figure 23. Between fifty thousand and ten thousand years ago, hunter-gatherer groups diversified and spread northeast toward the Americas and southeast toward Australia. By nine thousand years ago, two very divergent populations from this initial radiation—one centered on the northern Yellow River and one on the Yangtze River—independently developed agriculture, and then by five thousand years ago spread in all directions. In China, their collision created the gradient of northern and southern ancestry seen in the Han today.

The peoples of the Tibetan Plateau—who harbor a mixture of about two-thirds of their ancestry from the same Yellow River ghost population that contributed to the Han—are one example of this expansion. They likely brought farming for the first time to the region, as well as about one-third of their ancestry from an early branch of East Asians that plausibly corresponds to Tibet's indigenous hunter-gatherers.[36]

Another example is the Japanese. For many tens of thousands of years, the Japanese archipelago was dominated by hunter-gatherers, but after around twenty-three hundred years ago, mainland-derived agriculture began to be practiced and was associated with an archaeological culture with clear similarities to contemporary cultures on the Korean peninsula. The genetic data confirm that the spread of farming to the islands was mediated by migration. Modeling present-day Japanese as a mixture of two anciently divergent populations of entirely East Asian origin—one related to present-day Koreans and one related to the Ainu who today are restricted to the northernmost Japanese island and whose DNA is similar to that of pre-farming hunter-gatherers[37]—Naruya Saitou and colleagues estimated that present-day Japanese have about 80 percent farmer and 20 percent hunter-gatherer ancestry. Relying on the sizes of segments of farmer-related ancestry in present-day Japanese, we and Saitou estimated the average date of mixture to be around sixteen hundred years ago.[38] This date is far later than the first arrival of farmers to the region and suggests that after their arrival, it may have taken hundreds of years for social segregation between hunter-gatherers and farmers to break down. The date corresponds to the Kofun period, the first time when many Japanese islands were united under a single rule, perhaps marking the beginnings of the homogeneity that characterizes much of Japan today.

Ancient DNA is also revealing the deep history of humans in mainland Southeast Asia. In 2017, my laboratory extracted DNA from ancient humans at the almost four-thousand-year-old site of Man Bac in Vietnam, where people with skeletons similar in shape to those of Yangtze River agriculturalists and East Asians today were buried side by side with individuals with skeletons more similar to those of the previously resident hunter-gatherers.[39] Mark Lipson in my laboratory showed that in ancient Vietnam, all the samples we analyzed were a mixture of an early splitting lineage of East Eurasians and the Yangtze River Ghost Population, with the proportion of the Yangtze River Ghost Population higher in some of the Man Bac farmers we analyzed than in others. The main group of Man Bac farmers also had proportions of ancestry from these two lineages that were similar to those seen in present-day speakers of isolated Austroasiatic languages. These findings are consistent with the theory that Austroasiatic languages were spread by a movement of rice farmers from southern China who interbred with local hunter-gatherers.[40] Even today, large Austroasiatic-speaking populations in Cambodia and Vietnam harbor substantial albeit smaller proportions of this hunter-gatherer ancestry.

The genetic impact of the population spread that also dispersed Austroasiatic languages went beyond places where these languages are spoken today. In another study, Lipson showed that in western Indonesia where Austronesian languages are predominant, a substantial share of the ancestry comes from a population that derives from the same lineages as some Austroasiatic speakers on the mainland.[41] Lipson's discovery suggested that Austroasiatic speakers may have come first to western Indonesia, followed by Austronesian speakers with very different ancestry. This might explain why linguists Alexander Adelaar and Roger Blench noticed the presence of Austroasiatic loan words (words with an origin in another language group) in the Austronesian languages spoken on the island of Borneo.[42] Alternatively, Lipson's findings could be explained if Austronesian-speaking farmers took a detour through the mainland, mixing with local Austroasiatic-speaking populations there before spreading farther to western Indonesia.

The most impressive example of the movements of farmers from the East Asian heartland to the periphery is the Austronesian expansion. Today, Austronesian languages are spread across a vast region including hundreds of remote Pacific islands. Archaeological, linguistic, and genetic data taken together have suggested that around five thousand years ago, mainland East Asian farming spread to Taiwan, where the deepest branches of the Austronesian language family are found. These farmers spread southward to the Philippines about four thousand years ago, and farther south around the large island of New Guinea and into the smaller islands to its east.[43] At about the time they spread from Taiwan they probably invented outrigger canoes, boats with logs propped on the side that increase their stability in rough waters, making it possible to navigate the open seas. After thirty-three hundred years ago, ancient peoples making pottery in a style called Lapita appeared just to the east of New Guinea and soon afterward started expanding farther into the Pacific, quickly reaching Vanuatu three thousand kilometers from New Guinea. It took only a few hundred more years for them to spread through the western Polynesian islands including Tonga and Samoa, and then, after a long pause lasting until around twelve hundred years ago, they spread to the last habitable Pacific islands of New Zealand, Hawaii, and Easter Island by eight hundred years ago. The Austronesian expansion to the west was equally impressive, reaching Madagascar off the coast of Africa nine thousand kilometers to the west of the Philippines at least thirteen hundred years ago, and explaining why almost all Indonesians today as well as people from Madagascar speak Austronesian languages.[44]

Mark Lipson in my laboratory identified a genetic tracer dye for the Austronesian expansion—a type of ancestry that is nearly always present in peoples who today speak Austronesian languages. Lipson found that nearly all people who speak these languages harbor at least part of their ancestry from a population that is more closely related to aboriginal Taiwanese than it is to any mainland East Asian population. This supports the theory of an expansion from the region of Taiwan.[45]

Although there are genetic, linguistic, and archaeological common threads that make a compelling case for the Austronesian expansion, some geneticists balked at the suggestion that the first humans who peopled the remote islands of the Southwest Pacific during the Lapita dispersal were unmixed descendants of farmers from Taiwan.[46] How could these migrants have passed over the region of Papua New Guinea, occupied for more than forty thousand years, while mixing little with its inhabitants? Such a scenario seemed improbable in light of the fact that today, all Pacific islanders east of Papua New Guinea have at least 25 percent Papuan ancestry and up to around 90 percent.[47] How could this fit with the prevailing hypothesis that the Lapita archaeological culture was forged during a period of intense exchange between people ultimately originating in the farming center of China (via Taiwan) and New Guineas?

In 2016, ancient DNA struck again, disproving the view that had prevailed until then in the genetic literature. Well-preserved ancient DNA is hard to find in tropical climates like those of the southern Pacific. But the ability to get working DNA from the Pacific changed when, as described earlier, Ron Pinhasi and colleagues showed that DNA from the dense petrous bone of the skull containing the structures of the inner ear sometimes preserves up to one hundred times more DNA than can typically be obtained from other bones.[48] We initially struggled to study samples from the Pacific, but when we tried petrous bones, our luck changed.[49]

We succeeded at getting DNA from ancient people associated with the Lapita pottery culture in the Pacific islands of Vanuatu and Tonga who lived from around three thousand to twenty-five hundred years ago. Far from having substantial proportions of Papuan ancestry, we found that in fact they had little or none.[50] This showed that there must have been a later major migration from the New Guinea region into the remote Pacific. The late migration must have begun by at least twenty-four hundred years ago, as all the Vanuatu samples we have analyzed from that time and afterward had at least 90 percent Papuan ancestry.[51] How this later wave could have so comprehensively replaced the descendants of the original people who made Lapita pottery and yet retained the languages these people probably spoke remains a mystery. But the genetic data show that this is what happened. This is the kind of result that only genetics can deliver—the definitive documentation that major movements of people occurred. This proof of interaction between highly divergent peoples puts the ball back into the court of archaeologists to explain the nature and effects of those migrations.

Ancient DNA from the Southwest Pacific has continued to produce unexpected findings. When we and Johannes Krause's laboratory, working independently, analyzed the Papuan ancestry in Vanuatu, we found that it was more closely related to that in groups currently living in the Bismarck Islands near New Guinea than to

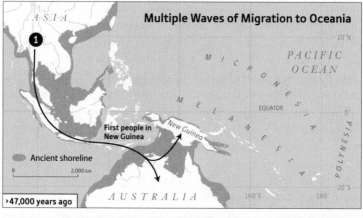

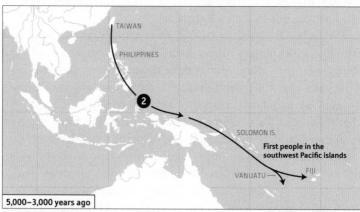

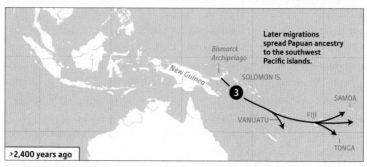

Figure 24. Ancient DNA shows that the first people of the southwest Pacific islands had none of the Papuan ancestry ubiquitous in the region today and that first arrived in New Guinea after fifty thousand years ago (top). The pioneer migrants had almost entirely East Asian ancestry (middle), and multiple later streams of migration brought primarily Papuan ancestry (bottom).

groups currently living in the Solomon Islands—despite the fact that the Solomon Islands are directly along the ocean sailing path to Vanuatu.[52] We also found that the Papuan ancestry present in remote Polynesian islands is not consistent with coming from the same source as that in Vanuatu. Thus there must have been not one, not two, but at least three major migrations into the open Pacific, with the first migration bringing East Asian ancestry and the Lapita pottery culture, and the later migrations bringing at least two different types of Papuan ancestry. So instead of a simple story, the spread of humans into the open Pacific was highly complex.

Can we ever hope to reconstruct the details of these migrations? There is every reason to be hopeful. Our picture of how the present-day populations of the Pacific islands formed is becoming increasingly clear thanks to access to ancient DNA from the region, and the fact that some islands have less complex population histories than mainland groups because of their isolation, permitting easier reconstruction of what occurred. Through genome-wide studies of modern and ancient populations, we will soon have a far more accurate picture of how humans moved through this vast region.

But right now our understanding of what happened in mainland East Asia remains murky and limited. The extraordinary expansion of the Han over the last two thousand years has added one more level of massive mixing to the already complex population structure that must have been established after thousands of years of agriculture in the region, and after the rise and fall of various Stone Age, Copper Age, Bronze Age, and Iron Age groups. This means that any attempt to reconstruct the deep population history of East Asia based on patterns of variation in present-day people must be viewed with great caution.

But as I write this chapter, the tsunami of the ancient DNA revolution is cresting, and it will shortly crash on East Asian shores. State-of-the-art ancient DNA laboratories have been founded in China, and are turning their powers to investigating collections of skeletal material that have been assembled over decades. Ancient DNA studies of these and other skeletons will reconstruct how the peoples of each ancient mainland East Asian culture relate to each other and to people living today. Our understanding of the deep interrelationships of the ancestral populations of East Asians, and of movements of people since the end of the last ice age, will soon be as clear as our understanding of what happened in Europe.

But it is hard to predict what ancient DNA studies in East Asia will show. While we are beginning to have a relatively good idea of what happened in Europe, Europe does not provide a good road map for what to expect for East Asia because it was peripheral to some of the great economic and technological advances of the last ten thousand years, whereas China was at the center of changes like the local invention of agriculture. What this means is that while we can be sure that the findings from ancient DNA studies in East Asia will be illuminating, we do not yet know what they will be. All we can be sure of is that ancient DNA studies will change our understanding of the human past in this most populous part of the world.

A Series of African Migrations

770,000–550,000 ya
Genetic estimate
of population separation
between Neanderthals
and modern humans

300,000–250,000 ya
*Middle Paleolithic
Transition*

70,000–50,000 ya
*Upper Paleolithic
Transition*

330,000–300,000 ya
Oldest fossils with features shared
with anatomically modern humans
(Jebel Irhoud, Morocco)

800,000
years ago

PERIOD OF
DETAIL

~700,000 ya
Possible split of an archaic African lineage that remixed
into Africans in the last tens of thousands of years

4,000–1,000 ya
Bantu expansion
spreads farming from
west-central Africa
and herding from
northeastern Africa.

1,800–800 ya
Pastoralist
expansion
spreads
Khoe-Kwadi
languages into
southern Africa.

6,000 ya   4,000   2,000   Present

8,100–1,400 ya
San ancestry,
now restricted
to South Africa,
was spread far
to the north.

›1,400 ya
Population
persists in
the islands
off Tanzania
with one-third
San-related
ancestry.

9,000 years ago – present

Rejoining Africa to the Human Story

A New Perspective on Our African Homeland

The recognition that Africa is central to the human story has, paradoxically, distracted attention from the last fifty thousand years of its prehistory. The intensive study of what happened in Africa before fifty thousand years ago is motivated by a universal recognition of the importance of the Middle to Later Stone Age transition in Africa and the Middle to Upper Paleolithic transition at the doorstep of Africa, those great leaps forward in recognizably modern human behavior attested to in the archaeological record. However, scholars have shown limited interest in Africa after this period. When I go to talks, a common slip of the tongue is that "we left Africa," as if the protagonists of the modern human story must be followed to Eurasia. The mistaken impression is that once Africa gave birth to the ancestral population of non-Africans, the African story ended, and the people who remained on the continent were static relics of the past, jettisoned from the main plot, unchanging over the last fifty thousand years.

The contrast between the richness of the information we currently have about the human story in Eurasia over the last fifty thousand years and the dearth of information about Africa over the same period is extraordinary. In Europe, where most of the research has been done, archaeologists have documented a detailed series of cultural transformations: from Neanderthals to pre-Aurignacian modern humans, to Aurignacians, to Gravettians, to the people who practiced Mesolithic culture, and then to Stone Age farmers and their successors in the Copper, Bronze, and Iron ages. The ancient DNA revolution—which has disproportionately sampled bones from Eurasia and especially from Europe—has further widened the gap in our understanding of the prehistory of Africa compared to that of Eurasia.

But of course, what all investigations that scratch below the surface show is that the people "left behind" in Africa changed just as much as the descendants of the people who emigrated. The main reason we don't know as much about the modern human story in Africa is lack of research. Human history over the last tens of thousands of years in Africa is an integral part of the story of our species. Focusing on Africa as the place where our species originated, while it might seem to highlight the importance of Africa, paradoxically does Africa a disservice by drawing attention away from the question of how populations that remained in Africa got to be the way they are today. With ancient and modern DNA, we can rectify this.

The Deep Mixture That Formed Modern Humans

In 2012, Sarah Tishkoff and her colleagues studied the biological impact of archaic admixture on the genomes of present-day Africans without access to ancient genomes like those of Neanderthals and Denisovans that had been used to document interbreeding between archaic and modern humans in Eurasia.[1]

Tishkoff and her colleagues sequenced genomes from some of the most diverse populations of Africa and analyzed their data to search for a pattern that is predicted when there has been interbreeding with archaic humans: very long stretches of DNA that have a high density of differences compared to the great majority of other genomes, consistent with an origin in a highly divergent population that was isolated until recently from modern humans.[2] When they applied this approach to present-day non-Africans, they pulled out stretches of DNA that they found were nearly exact matches to the Neander-thal sequence. Tishkoff and her colleagues also found long stretches of deeply divergent sequences in present-day Africans whose ances-tors did not mix with Neanderthals. Since Neanderthals have con-tributed little if any ancestry to Africans, this was likely to have been the result of mixture with mystery African archaic humans—ghost populations whose genomes have not yet been sequenced.

Jeffrey Wall and Michael Hammer, using the same types of genetic signatures, attempted to learn something about the relationship of the archaic populations to present-day Africans.[3] They estimated that the archaic population separated from the ancestors of present-day humans in Africa about seven hundred thousand years ago and remixed around thirty-five thousand years ago, contributing about 2 percent of the ancestry of some present-day African populations. However, it is important to view these dates and estimated propor-tions of mixture with caution because of uncertainties about the rate at which mutations occur in humans and because of the limited amount of data Wall and Hammer analyzed.

The possibility of admixture between modern and archaic humans in sub-Saharan Africa is exciting, and there are even human remains from West Africa dating to as late as eleven thousand years ago with archaic features, providing skeletal evidence in support of the idea that archaic and modern human populations coexisted in Africa until relatively recently.[4] Thus, there were ample opportunities for interbreeding with archaic humans as modern humans expanded in Africa, just as in Eurasia.

If the proportion of admixture with archaic African humans was only around 2 percent as Wall and Hammer estimated, it is likely to have had only a modest biological effect, similar to the effect of the contribution of Neanderthals and Denisovans to the genetic makeup of present-day people outside Africa. However, this does not rule out the possibility of major mixture events in deep African history. The best evidence for deep mixture of modern human populations in sub-Saharan Africa comes from the frequencies of mutations. One generation after a mutation occurs, it is extremely rare as it is present in only a single person. In subsequent generations, the mutation's frequency fluctuates upward or downward at random, depending on the number of offspring to which it happens to be transmitted. Most mutations never achieve a substantial frequency, as at some point the few individuals carrying them happen not to transmit them to their children, causing them to fluctuate down to 0 percent frequency and disappear forever.

The effect of this constant pumping into the population of rare new mutations is that there are expected to be fewer common than rare mutations in a population. The frequencies of mutations that are variable in a population are in fact expected to follow an inverse law, with twice as many mutations that occur at 10 percent frequency as those that occur at 20 percent frequency, and twice as many of these in turn as those that occur at 40 percent frequency.

My colleague Nick Patterson tested this expectation, focusing on mutations present in a large sample of individuals from the Yoruba group of Nigeria that were also present in the Neanderthal genome.[5] Patterson's focus on mutations present in Neanderthals was clever; he knew that mutations discovered in this way were almost certainly frequent in the common ancestral population of humans and Neanderthals, and by implication in their descendants too. Mathematically, the expectation that such mutations would be common is exactly counterbalanced by the inverse law, with the result that mutations meeting these criteria are expected to be equally distributed across all frequencies.

But the real data showed a different pattern. When Patterson examined sequences from present-day Yoruba, he observed a greatly elevated rate of mutations both at very high and low frequencies, instead of an equal distribution across all frequencies. This "U-shaped" distribution of mutation frequencies is what would be expected in the case of ancient mixture. After two populations separate, random frequency fluctuation occurs in each population, so that the mutations that fluctuate by chance to 0 percent or 100 percent frequency in one population are by and large not expected to be the same as those that do so in the other population. When the populations then remix, the mutations that rose to extreme frequencies in one population but not in the other would be reintroduced as variable genetic types. This would produce peaks of extra mutation density in the mixed population. The first peak corresponding to mutations that rose to extreme frequencies in the first population is expected to start at the proportion of mixture, while the second peak corresponding to mutations that rose to extreme frequencies in the second population is expected to start at 100 percent minus the proportion of mixture. This is exactly the pattern that Patterson found, and he showed that it could be explained if Yoruba descended from a mixture of two highly differentiated human populations in close to equal proportions.

Patterson tested whether the patterns he observed were consistent with a model in which only Yoruba descend from this mixture but non-Africans do not. But this was contradicted by the data. Instead, all non-Africans—and even divergent African lineages such as San hunter-gatherers—also seem to be descended from a similar mixture. Thus, although Patterson had begun by looking at West Africans, the mixture event he detected was not specific to that population. Rather, it seemed to be a shared event in the ancestry of present-day humans, suggesting that the mixture may have occurred close to the time when anatomically modern human features first appear in the skeletal record after around three hundred thousand years ago.[6]

Patterson's findings resonated with a discovery from the 2011 study by Heng Li and Richard Durbin (discussed in part I) that reconstructed human population size history from a single person's genome.[7] That study compared the genome sequence a person gets from his or her mother to the sequence he or she gets from his or her father. It found fewer locations in the genome where the reconstructed age of the shared ancestor falls between 400,000 and 150,000 years ago than would be expected if the population had been constant in size.[8] One possible explanation for this result is that the ancestral population of all modern humans was very large over this period, which would mean that the probability that any two genomes today share a particular ancestor at this time is small (there being many possible ancestors in each generation). But the other possibility was that the ancestral human population consisted of multiple highly divergent groups instead of a single, freely mixing group, and hence the lineages ancestral to present-day people were isolated in separate populations at this time. This pattern could be reflecting the same mixture event that Patterson had highlighted through his study of mutation frequencies. The reconstructed time corresponds to a period when there is skeletal evidence of archaic human forms overlapping with modern human forms in Africa. For example, the *Homo naledi* skeletons recently discovered in a cave in South Africa had relatively modern human bodies but brains much smaller than those of modern humans, and date to between 340,000 and 230,000 years ago.[9]

There was also a third line of evidence for archaic mixture. A commonly held view is that the San hunter-gatherers of southern Africa largely derive from a lineage that branched off the one leading to all other present-day modern human lineages before they separated from one another.[10] If so, the San would be expected to share mutations at exactly the same rate with all non–southern Africans. But Pontus Skoglund in my laboratory showed that the San share more mutations with eastern and central African hunter-gatherers than they do with West African populations like the Yoruba of Nigeria.[11] This could be explained if the West African populations harbor more ancestry from one of the early-splitting populations than is the case for non-African populations. Perhaps all present-day humans are a mixture of two highly divergent ancestral groups, with the largest proportion in West Africans, but all populations inheriting DNA from both.

These results suggest the possibility that major mixture in Africa occurred in the time well before around fifty thousand years ago when modern human behavior burst into full flower in the archaeological record. This mixture wasn't a minor event, such as the approximately 2 percent Neanderthal admixture in non-Africans or the ghost archaic ancestry in Africans found by Wall and Hammer. Because this mixture was closer to 50/50, it is not even clear which one of the source populations should properly be considered archaic and which modern. Perhaps neither was modern, or neither was archaic. Perhaps the mixture itself was essential to forging modern humans, bringing together biological traits from the two mixing populations and combining them in new ways that were advantageous to the newly formed populations.

A Possible New Deep Lineage in the Ancestry of Modern Humans

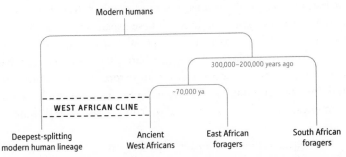

Figure 25. The deep relationships among present-day modern human lineages are far from simple. One model that genomic findings suggest is that the oldest modern human split in Africa led to a lineage represented in highest proportion in West Africa, a split that must have occurred before three hundred thousand to two hundred thousand years ago, the date of the separation of ancient East and South African foragers. An expansion of modern humans associated with the Later Stone Age and Upper Paleolithic transitions after around fifty thousand years ago could then have connected all populations in Africa.

How Agriculture Threw a Veil over Africa's Past

How can we begin to learn what happened in Africa after the ancestral population of modern humans was forged, and also after the ancestors of present-day non-Africans spread out of Africa and the Near East beginning around fifty thousand years ago? There is a lot of information to work with, as African genome sequences are typically about a third more diverse than non-African ones. Human diversity in Africa is extraordinary not only within but also across populations, as some pairs of African populations have been isolated for up to four times longer than any pairs of populations outside the continent, as reflected by the fact that for some pairs of populations—like San hunter-gatherers from southern Africa and Yoruba from West Africa—the minimum density of mutations separating their genomes is that much greater than that for any pair of genomes outside Africa.[12]

But learning about Africans' deep past from today's populations is extremely challenging because while much of the ancient variation still exists in people living today, it is all mixed up. The most recent mixing of populations occurred in the last few thousand years due to at least four great expansions, all of which are associated with the spread of language groups, and most of which have been driven by the spread of agriculturalists.[13] These expansions have thrown a veil over the African past, moving populations thousands of kilometers from their places of origin, where they displaced or mixed with populations that were widespread before. In this respect, the study of African populations is no different from the study of Eurasian ones, which have also turned over in the last several thousand years.

The agriculturalist expansion that had the greatest impact on Africa is the one associated with people who speak languages of the Bantu family.[14] Archaeological studies have documented how beginning around four thousand years ago, a new culture spread out of the region at the border of Nigeria and Cameroon in west-central Africa. People from this culture lived at the boundary of the forest and expanding savanna and developed a highly productive set of crops that was capable of supporting dense populations.[15] By about twenty-five hundred years ago they had spread as far as Lake Victoria in eastern Africa and mastered iron toolmaking technology,[16] and by around seventeen hundred years ago they had reached southern Africa.[17] The consequence of this expansion is that the great majority of people in eastern, central, and southern Africa speak Bantu languages, which are most diverse today in present-day Cameroon, consistent with the theory that proto-Bantu languages originated there and were spread by the culture that also expanded from there around four thousand years ago.[18] Bantu languages are a subset of the larger Niger-Kordofanian family spanning most of the languages of West Africa,[19] which likely explains why today the frequencies of mutations in groups in Nigeria and in Zambia are more similar than the frequencies of mutations in Germany and Italy despite the former two countries being separated by a far greater geographic distance.

Ultra-sensitive genetic methods, which can detect shared relatives of pairs of individuals in the past few thousand years, have now made it possible to learn something about the geographic path of the

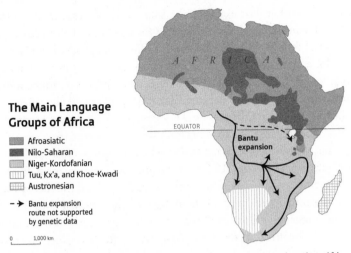

**The Main Language
Groups of Africa**

- Afroasiatic
- Nilo-Saharan
- Niger-Kordofanian
- Tuu, Kx'a, and Khoe-Kwadi
- Austronesian

– ➤ Bantu expansion
route not supported
by genetic data

0   1,000 km

Figure 26. Today, West African–related ancestry is predominant in eastern and southern Africa due to the Bantu expansion of the last four thousand years.

Bantu expansion. The genetic variation in Bantu speakers in East Africa is more closely related to the genetic variation in Malawi to the south of the Central African rainforests than it is to genetic variation in Cameroon.[20] This suggests that the initial Bantu expansion was largely to the south and that the movement to East Africa was a later expansion from a southern staging ground. This contrasts with the theory of a direct eastward movement from Cameroon, a theory that had been plausible prior to the genetic data.

Another agricultural expansion that had a profound impact is the one that spread Nilo-Saharan languages, spoken by groups from Mali to Tanzania. Many Nilo-Saharan speakers are cattle herders, and a common view is that the Nilo-Saharan expansion was driven by the spread of farming and herding in Africa's dry Sahel region during the expansion of the Sahara Desert over the last five thousand years. One important branch of Nilo-Saharan is the Nilotic languages, which are mostly spoken by cattle herders along the Nile River and in East Africa, including the Maasai and Dinka. The genetic data make it clear that Nilotic-speaking herders were not always socially disadvantaged relative to farmers in the frontier regions where they encountered each other. For example, the Luo group of western Kenya (to which former U.S. president Barack Obama's father belonged) are a primarily farming people who speak a Nilotic language. But George Ayodo, a Luo scientist from Kenya who spent time in my laboratory, found that the mutation frequencies in the Luo are much more similar to those of the majority of Bantu speakers, likely reflecting a history in which a Bantu-speaking group in East Africa adopted a Nilo-Saharan language from its high status neighbors.[21]

The African language expansion whose origin is most unclear is the one associated with Afroasiatic languages. They are most diverse in present-day Ethiopia, which throws weight behind the theory that northeastern Africa was the homeland of the original speakers of these languages.[22] But the Afroasiatic language family also contains a branch localized to the Near East that includes Arabic, Hebrew, and ancient Akkadian. It has been hypothesized on this basis that the spread of Afroasiatic languages, or at least some branches of them, could have been related to the spread of Near Eastern agriculture,[23] which introduced barley and wheat and other Near Eastern crops into northeast Africa up to seven thousand years ago.[24] New insights are already emerging from ancient DNA, which makes it possible to document ancient migrations between the Near East and North Africa that could have spread languages, culture, and crops. In 2016 and 2017, my laboratory published two papers showing that a shared feature of many East African groups, including ones that do not speak Afroasiatic languages, is that they harbor substantial ancestry from people related to farmers who lived in the Near East around ten thousand years ago.[25] Our work also found strong evidence for a second wave of West Eurasian–related admixture—this time with a contribution from Iranian-related farmers as might be expected from a spread from the Near East in the Bronze Age—and showed that this ancestry is widespread in present-day people from Somalia and Ethiopia who speak Afroasiatic languages in the Cushitic subfamily. So the genetic data provide evidence for at least two major waves of north-to-south population movement in the period when Afroasiatic languages were spreading and diversifying, and no evidence of south-to-north migration (there is little if any sub-Saharan African related ancestry in ancient Near Easterners or Egyptians prior to medieval times).[26] Genes do not determine what language a person speaks and so genetic data cannot by themselves determine how languages spread, and thus cannot provide definitive evidence in favor of one theory or another about whether the ultimate homeland of Afroasiatic languages was sub-Saharan Africa, North Africa, Arabia, or the Near East. But there is no question that the genetic data increase the plausibility of a Near Eastern agriculturalist source for at least some Afroasiatic languages, and the genetic findings raise the question of what languages were spoken by these north-to-south migrants.

The fourth great agriculturalist expansion in Africa is the one associated with the Khoe-Kwadi languages of southern Africa. Like the two language groups spoken by hunter-gatherer groups in the south of Africa—Kx'a and Tuu—Khoe-Kwadi languages are characterized by click sounds. Based on shared words for herding, it has been hypothesized that Khoe-Kwadi languages were brought from East Africa by cattle herders who came to southern Africa after eighteen hundred years ago and who may also have picked up click sounds from local populations.[27] The genetic data support the hypothesis of a major genetic contribution by East Africans to Khoe-Kwadi-speaking populations today. In 2012, Joseph Pickrell in my laboratory showed that Khoe-Kwadi speakers share a disproportionate amount of their ancestry with Ethiopians compared to the Kx'a and the Tuu, as might be expected from a migration from the north.[28] The size of East African–derived DNA segments in some of the Khoe-Kwadi–speaking populations is what would be expected from mixture eighteen to nine hundred years ago with a ghost herder population, consistent with the arrival of herders around this time and a delay before the mixture with local populations was complete. Within the segments matching East Africans, Pickrell found even smaller segments that matched Near Easterners more than they did any other populations, and that had lengths expected for an average mixture date of around three thousand years ago. That is the average date of mixture between people of West Eurasian–related ancestry and sub-Saharan ancestry in many groups in Ethiopia,[29] so this finding provides further support for the hypothesis of an East African source.

Ancient DNA has now verified this hypothesis. In 2017, Pontus Skoglund analyzed ancient DNA from the approximately thirty-one-hundred-year-old remains of an infant girl from Tanzania in equatorial East Africa, and an approximately twelve-hundred-year-old sample from the western Cape region of South Africa, both buried among artifacts and animal bones that identified them as being from herder populations.[30] The Tanzanian girl was a member of the ghost herding population that Pickrell and I had predicted: a group that derived most of its ancestry from ancient East African hunter-gatherers, and the remaining part from an ancient West Eurasian–related population. This population almost certainly played a major role in spreading cattle herding from the Near East and North Africa across sub-Saharan Africa. Our ancient DNA evidence from the southern African herder also strongly supported this idea, showing that this individual derived about one-third of her ancestry from the pastoralist population of which the Tanzanian girl was a part, and her remaining ancestry from local groups related to present-day San hunter-gatherers. The mixture of ancestries in the twelve-hundred-year-old southern African herder was very similar to that in present-day Khoe-Kwadi speakers, many of whom are herders, supporting the theory that early Khoe-Kwadi languages, herding, and this type of East African ancestry all spread to southern Africa through a movement of people.

The landscape of human biological and cultural diversity in Africa today, dominated as it is by the effects of the agricultural expansions of the last few thousand years, is extraordinary, but it is also distracting if one's interest is in understanding the big picture of what happened. A trap that researchers of African genetics, archaeology, and linguistics repeatedly fall into is celebrating Africa's present-day diversity, epitomized by a slide showing the faces of people from across the continent who look very different from each other that many of us use when presenting on Africa. It is tempting to think that in order to comprehend deep time in Africa we need to be able to hold all of that diversity in our heads and explain all of it at once. But most of the present-day population structure of Africa is shaped by the agricultural expansions of the past few thousand years, and so focusing on describing Africa's mesmerizing diversity paradoxically does the project of understanding the big picture of humans in Africa a disservice just as much as focusing on the common origins of all modern humans in Africa does Africa a disservice. We need to stop focusing on describing the veil and instead rip it away, and for this we need ancient DNA.

Reconstructing Africa's Forager Past

Who lived in Africa before the expansion of food producers, the people who so profoundly transformed the human landscape of the continent? Answering this question is extraordinarily difficult based on patterns of present-day variation. In the introduction to this book, I described how Luca Cavalli-Sforza made a bet in 1960 that it would be possible to reconstruct the deep history of human populations based entirely on patterns of genetic variation in present-day groups.[31] However, he lost his bet, as ancient DNA has revealed that there has been so much migration and population extinction that in most instances it is very difficult even with sophisticated statistical methods to recover the details of ancient demographic events from the traces left behind in the DNA of present-day people.

The breakthrough that is making it possible to get beyond this impasse will not be surprising to the reader. It is genome-wide ancient DNA, which can be coanalyzed with data from groups that have been genetically and culturally isolated compared to their neighbors, among them the Pygmies of Central Africa, the San hunter-gatherers of the southern tip of Africa, and the Hadza of Tanzania, whose languages with clicks are very different from the languages of the Bantu who surround them and whose genetic ancestry is highly distinctive as well. Some of these populations harbor genetic lineages that are highly divergent from their neighbors. We can compare data from these ancient samples to probe events that occurred deeper in time than those that can be accessed only by analyzing the DNA of present-day populations.

Well-preserved ancient DNA has until recently been hard to find in most parts of Africa because of the hot climate, which accelerates chemical reactions that degrade DNA. But in 2015, the ancient

DNA revolution finally arrived in Africa because of improvements in the efficiency of DNA extraction techniques and a better understanding of which bones yielded the most DNA.

The first genome-wide ancient DNA data from Africa came from a forty-five-hundred-year-old skeleton found in a highland cave in Ethiopia.[32] This ancient individual was much more closely related to one group living in Ethiopia today, the Ari, than to many others. Today there is an intricate caste system that shapes the lives of many people within Ethiopia, with elaborate rules preventing marriage between groups with different traditional roles.[33] The Ari include three subgroups—the Cultivators, Blacksmiths, and Potters—who are socially and genetically differentiated from one another and from non-Ari groups.[34] Since the Ari have a distinctive genetic affinity to the forty-five-hundred-year-old ancient highland individual compared to other Ethiopian groups, it is clear that there were strong local barriers to gene exchange and homogenization within the region of present-day Ethiopia that persisted for at least forty-five hundred years. This is the best example of strong endogamy that I know of—even more ancient than the evidence of endogamy in India that so far is only documented as going back a couple of thousand years.[35]

Ancient DNA keeps surprising us. In 2017, Pontus Skoglund in my laboratory analyzed sixteen individuals from Africa: foragers and herders from South Africa who lived between about twenty-one hundred and twelve hundred years ago, foragers from Malawi in southern Africa who lived between about eighty-one hundred and twenty-five hundred years ago, and foragers, farmers, and herders from Tanzania and Kenya who lived between about thirty-one hundred and four hundred years ago.[36] While these individuals are very recent compared to some of the oldest Eurasian ancient DNA, they nevertheless provide insights into African population structure before the arrival of the food producers who transformed much of Africa's human geography.

A great surprise that emerged from our ancient DNA analysis was that there was evidence of a ghost population dominating the eastern seaboard of sub-Saharan Africa that appears to have been largely displaced by the expansion of agriculturalists.[37] This population, which we called the "East African Foragers," contributed all of the ancestry of two ancient hunter-gatherer genomes in our dataset from Ethiopia and Kenya, as well as essentially all of the ancestry of the present-day Hadza of Tanzania, who today number fewer than one thousand. We also found that the East African Foragers were more closely related to non-Africans today than they were to any other groups in sub-Saharan Africa. The close relationship to non-Africans suggests that the ancestors of the East African Foragers may have been the population in which the Middle to Later Stone Age transition occurred, propelling expansions outside of Africa and possibly within Africa too after around fifty thousand years ago. So the population that became the East African Foragers had a pivotal role in our history.

The East African Foragers were not a homogeneous population. This is evident from the fact that our data include at least three distinct East African Forager groups within Africa—one spanning the ancient Ethiopian and ancient Kenyan, a second contributing large fractions of the ancestry of the ancient foragers from the Zanzibar Archipelago and Malawi, and a third represented in the present-day Hadza.[38] Based on the sparse data we had, we were not able to determine the date when these groups separated from one another. But given the extended geographic span and the antiquity of human occupation in this region, it would not be surprising if some of the differences among these groups dated back tens of thousands of years. There is precedent for such separations within forager populations in Africa. In 2012, my laboratory and another showed that a group that I think of as the "South African Foragers"—a lineage that is as divergent from the East African Foragers as any present-day human population—contained within it two highly divergent lineages that separated from each other at least twenty thousand years ago.[39] East Africa is at least as rich a human habitat as southern Africa, and it would not be surprising if separations among foragers in East Africa were at least as old.

The second surprise was our discovery that some of our ancient African forager population samples shared ancestry from both South African Forager lineages and East African Forager lineages. Today, South African Forager lineages are essentially entirely restricted

The Pre-Bantu Population Structure of Eastern Africa

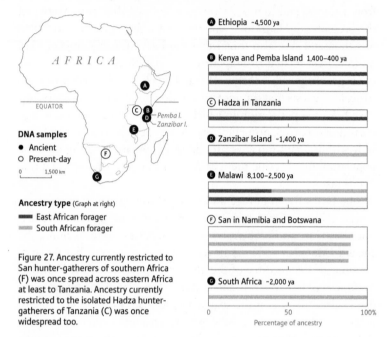

DNA samples
- Ancient
○ Present-day

0   1,500 km

Ancestry type (Graph at right)
▬▬ East African forager
▭▭ South African forager

Figure 27. Ancestry currently restricted to San hunter-gatherers of southern Africa (F) was once spread across eastern Africa at least to Tanzania. Ancestry currently restricted to the isolated Hadza hunter-gatherers of Tanzania (C) was once widespread too.

to southernmost Africa, where they form an important part of the ancestry of nearly all of the populations that use languages with clicks in them, and where they contribute almost all the ancestry of present-day San foragers as well as the ancient forager genomes we generated from southern Africa. But our ancient samples show that the term "South African Forager" may be misleading about where the ancestral population of this group arose. Two approximately fourteen-hundred-year-old individuals from Zanzibar and Pemba islands off the coast of Tanzania—an island chain that separated from the mainland approximately ten thousand years ago as sea levels rose and thus plausibly harbors isolated descendants of a forager population that lived in East Africa around that time[40]—were a mixture of approximately one-third South African Forager–related ancestry and the remainder East African Forager ancestry.[41] A series of seven samples from three different archaeological sites in Malawi in south-central Africa, which we dated to between about eighty-one hundred and twenty-five hundred years ago, were part of a homogeneous population that harbored about two-thirds South African Forager–related ancestry and the remainder East African Forager ancestry. So South African Forager ancestry was in the past distributed over a much broader swath of the continent, making it hard to know where this ancient population originated.

Ancient DNA is teaching us that the history of modern Africa has its roots in ancient population separations and mixtures even before the arrival of agriculture. Thus the human story in Africa is complex at all levels and at all time depths, as might be expected from the continent's huge size, its varied landscape, and the antiquity of the presence of our species there. The ancient DNA revolution is only just getting a toehold in Africa. In the coming years, Africa will be fully included in the ancient DNA revolution, and data will arrive from remains from more locations and from deeper times. These data will surely transform and clarify our view of what happened in the deep African past.

What's Next for Understanding the African Story

Some of the most striking examples of the complexity of human population structure in Africa are the patterns of natural selection on the continent. People of West African ancestry today have a high rate of sickle cell disease, conferred by a mutation that changes the blood protein hemoglobin, the molecule that more than any other is responsible for ferrying oxygen around the body. This mutation has risen to substantial frequency under the pressure of natural selection in several places in Africa: in far West Africa (e.g., Senegal), in west-central Africa (e.g., Nigeria), and in central Africa (whence the mutation spread to eastern Africa and southern Africa via the migrations associated with the Bantu expansion). The reason this mutation has risen to such a high frequency in each of these populations is that if a person carries one copy of the mutation from either of his or her parents, it protects against the infectious disease malaria. Malaria is so dangerous that the protection provided to the approximately 20 percent of the population who carry one copy of the sickle cell mutation is balanced in evolutionary terms with the cost that the approximately 1 percent of the population has to pay in carrying two copies of the mutation and suffering from sickle cell disease, which kills in childhood without treatment. Strikingly, the mutation has arisen independently in each of three locations in Africa, which we know from the fact that the sequences on which it resides are all different. From a naive perspective this seems surprising, as one would think that a mutation like this would be so advantageous to the people who carry it that once it arose it would spread around the vast malaria zone of Africa propelled by a tailwind of natural selection if there was even a small rate of interbreeding among neighbors.[42] A similar pattern is seen for the mutations in the lactase gene that confer an ability to digest cow's milk into adulthood. The genetic basis for lactase persistence is completely different in North Africans and in the Fulani of West Africa than it is in the Masai of Sudan and Kenya, who carry different mutations, albeit in the same gene.[43]

As Peter Ralph and Graham Coop have shown, the multiple origins in Africa of sickle cell mutations and of mutations that allow people to digest cow's milk imply that the rate of migration among these populations—even in parts of sub-Saharan Africa less than a couple of thousand kilometers from each other—has been extraordinarily low since the need for these mutations arose. As a result, the most efficient way for evolutionary forces to spread beneficial mutations has often been to invent mutations anew rather than to import them from other populations.[44] The limited migration rates between some regions of Africa over the last few thousand years has resulted in what Ralph and Coop have described as a "tessellated" pattern of population structure in Africa. Tessellation is a mathematical term for a landscape of tiles—regions of genetic homogeneity demarcated by sharp boundaries—that is expected to form when the process of homogenization due to gene exchanges among neighbors competes with the process of generating new advantageous variations in each region. The size of the regions where the same sickle cell mutation or same lactase-persistence mutation prevails reflects the rate of gene exchange among neighboring populations in Africa over the last thousands of years.

Our understanding of African population history is still in its early stages, but it is already clear that the story is complicated, with separations within major lineages such as East African Foragers and South African Foragers dating back deep in time, and layers of mixture beyond the most recent ones that have arisen due to the spread of agriculture. Eventually, by obtaining many more samples of ancient DNA from Africa, we will be able to comprehend the range of human variation in Africa in the last tens of thousands of years and make meaningful reconstructions of population structure.

What we can already be sure of is that in Africa, as in every region that has yielded ancient DNA, the model of an evolutionary tree in which today's populations have remained unchanged and separate since branching from a central trunk is dead, and that instead the truth has involved great cycles of population separation and mixture. What we can be sure of, too, is that in Africa, as in every world region that has yielded ancient DNA, the data will disprove many commonly held assumptions. The implications of this complexity for society, and for the way we need to rethink who we are, is the theme of part III of this book.

Part III

The Disruptive Genome

10

The Genomics of Inequality

The Great Mixing

The American melting pot began to swirl almost as soon as Christopher Columbus arrived in 1492. European colonists, their African slaves, and the indigenous Americans were from populations whose ancestors had been isolated from each other for tens of thousands of years. Within a few years of meeting they began mixing, founding new populations that today number in the hundreds of millions.

Martín Cortés "el Mestizo" belonged to one of the first of those populations. He was born within four years of the start of the 1519 military campaign in which his father, Hernán Cortés, led just five hundred soldiers to overthrow the Aztec Empire that dominated Mexico. His mother, "La Malinche," was one of twenty female captives given over to the Spanish after a battle and she first served as an interpreter before becoming the mistress of Hernán Cortés. The Spanish quickly invented a term for the people of combined European and Native American ancestry who emerged from unions like this. "Mestizo" comes from the Spanish word *mestizaje*, which in English means miscegenation—the mixing of different "racial" types. To maintain their status in the social hierarchy, the Spanish and Portuguese set up a *casta* system in which people of entirely European ancestry (especially those born in Europe) had the highest status, while people who had even some non-European ancestry had lower status. This system collapsed under the demographic inevitability of admixture; within a few centuries people of entirely European ancestry were either an extreme minority or gone, and it was no longer feasible to limit power to those with entirely European ancestry. Following the independence movements of the nineteenth and early twentieth centuries, mixed ancestry became a source of pride in South and Central America. In Mexico, it defines national identity.[1]

Migration of Africans to the Americas after 1492 occurred on a similar scale as migration of Europeans. All told, an estimated twelve million enslaved Africans were forced to make the journey, jammed into the holds of ships before being sold at auction.[2] Slave traders from Spain, Portugal, France, Britain, and the United States made great fortunes by satisfying the colonialists' need for manual labor. African slaves worked in the silver mines of Peru and Mexico and raised crops such as sugarcane and eventually tobacco and cotton. Africans were less affected by Old World diseases than Native Americans and easier to exploit than indigenous people, as they were far from home and scattered among a population that did not speak their languages. Deprived of their cultural points of reference, slaves had little ability to organize or resist. Most were sold in South America or the Caribbean, where they were often worked to death. Around 5 to 10 percent were brought to what became the United States. Following the first recorded sale of slaves by Portuguese traders in 1526, the rate of importation into the New World increased, plateauing at around seventy-five thousand per year until the trans-Atlantic slave trade was outlawed—in the British colonies in 1807, in the United States in 1808, and in Brazil in 1850.

Today there are hundreds of millions of people in the Americas with African ancestry, the largest numbers in Brazil, the Caribbean, and the United States. The mixing of three highly divergent populations in the Americas—Europeans, indigenous people, and sub-Saharan Africans—that began almost five hundred years ago continues to this day. Even in the United States, where European Americans are still in the majority, African Americans and Latinos comprise around a third of the population. Nearly all individuals from these mixed populations derive large stretches of their genomes from ancestors who lived on different continents fewer than twenty generations ago.

A small percentage of European Americans have large stretches of African or Native American DNA as well, the legacy of people who successfully "passed" themselves off into the white majority.[3]

A 1973 science-fiction novel, Piers Anthony's *Race Against Time*, envisions a future in which the mixing of populations initiated by European colonialism reaches its inevitable conclusion, and by the year 2300 nearly all humans belong to a "Standard" population.[4] In that year, only six unmixed people are left: one pair of "pure-bred Caucasians," one pair of "purebred Africans," and one pair of "purebred Chinese." These "purebreds" are being raised in human zoos by foster parents and are being groomed to breed with the only remaining individual of similar ancestry to sustain humanity's diversity, a diversity that is viewed by the "Standard" population as a resource of irreplaceable biological value on the verge of being lost. The premise of the novel is that the centuries after 1492 were a uniquely homogenizing time in the history of our species, a period of unprecedented mixing of previously separated populations enabled by transoceanic travel, which brought together groups whose ancestors had not been in contact with one another for tens or hundreds of thousands of years.

But this premise is mistaken. The genome revolution has shown that we are not living in particularly special times when viewed from the perspective of the great sweep of the human past. Mixtures of highly divergent groups have happened time and again, homogenizing populations just as divergent from one another as Europeans, Africans, and Native Americans. And in many of these great admixtures, a central theme has been the coupling of men with social power in one population and women from the other.

Founding Fathers

Not long after the Constitutional Convention of 1787, the man who would become the United States' third president, Thomas Jefferson, began a sexual relationship with his slave Sally Hemings. Jefferson owned a large plantation in the state of Virginia, where some 40 percent of the population was enslaved.[5] Sally Hemings was an African American slave with three European grandparents. But her mother's mother was a slave of African descent, and under Virginia law the status of a slave was maternally inherited. Jefferson and Hemings had six children together.[6]

The Jefferson-Hemings relationship has been disputed by some who have suggested that Jefferson—who is America's greatest Enlightenment thinker and the author of the U.S. Declaration of Independence—would not have maintained an illegitimate family. However, a genetic study published in 1998 revealed a Y-chromosome match between the male-line descendants of Eston Hemings Jefferson, the youngest son of Sally Hemings, and the male-line descendants of Jefferson's paternal uncle.[7] The genetic findings could in theory be explained if a male relative of Jefferson was the father rather than Jefferson himself. But there is no historical evidence for this possibility, and there is a credible nineteenth-century account of the Hemings-Jefferson relationship from Madison Hemings, another son of Hemings. A study by the Thomas Jefferson Memorial Foundation in 2000 concluded that, with high probability, the story was true.[8]

According to the account of Madison Hemings, his mother had a chance at freedom because she joined Jefferson in France, where slavery was illegal, but she agreed to return as a slave to the United States with Jefferson under the condition that their children would eventually be set free. Hemings was thirty years younger than Jefferson, and in France, where she began her relationship with him between the ages of fourteen and sixteen, she was dependent on him. She was also the half sister of Jefferson's wife, Martha Randolph, who had died of complications of childbirth several years earlier and whose father had a secret relationship with the mother of Sally Hemings.[9]

Historians have attempted to quantify how widespread families like these were in the United States. Mixed-ancestry unions were often unrecorded, and when they were, children were categorized in different ways by different states. Genetics can help here. Although so far no one has analyzed DNA from African American graveyards to chart the emergence of a mixed-ancestry community in the

United States, genetic studies of the present-day African American population are already enriching our understanding. Mark Shriver led a 2001 study that analyzed mutations that are extremely different in frequency between present-day Europeans and West Africans in order to study the African American populations of South Carolina. Shriver and his colleagues used these results to estimate the proportion of ancestors who lived in Europe a few dozen generations earlier.[10] The highest proportion, around 18 percent, is found in the inland state capital, Columbia, a percentage at the low end of the range of cities in other U.S. states. They estimated about 12 percent European ancestry along the South Carolina coast, including in the slave port of Charleston, which they thought might reflect waves of slave importation keeping the African ancestry high. They estimated the lowest proportion of European ancestry, around 4 percent, on the Sea Islands off the coast, reflecting the history of isolation of the slaves who settled there, an isolation attested to by the fact that the Sea Islanders are the only African Americans still speaking a language, Gullah, with an African-derived grammar. Comparison of Y-chromosome and mitochondrial DNA types that are highly different in frequency between African Americans and Europeans also shows that by far the majority of the European ancestry in these populations comes from males, the result of social inequality in which mixed-race couplings were primarily between free males and female slaves.[11]

The patterns in South Carolina are a microcosm of those in the United States as a whole. Katarzyna Bryc, at the personal ancestry testing company 23andMe, worked with me to analyze more than five thousand self-described African Americans in the company database, and found that the average European ancestry proportion was 27 percent in most of the genome but only 23 percent on chromosome X.[12] Comparing proportions of ancestry on chromosome X and the other chromosomes can provide information about differences in male and female behavior during population mixture, because two-thirds of X chromosomes in the world are carried in females compared to only about half of all other chromosomes, so the X chromosome is relatively more influenced by female history. By computing the proportion of European male and female ancestors that would be necessary to produce the observed difference in European ancestry between chromosome X and the autosomes, Bryc was able to estimate the separate male (38 percent) and female (10 percent) proportion of European ancestors in African Americans. These numbers imply that the contribution of European American men to the genetic makeup of the present-day African American population is about four times that of European American women. When I discussed these findings with the sociologist Orlando Patterson, he pointed out that the fraction of the European ancestry in African Americans that came from males—which if different from half is called "sex bias"—must have been far greater during the time of slavery. Since the civil rights movement in the United States in the mid-twentieth century, cultural changes have caused the sex bias to reverse, with more coupling between black men and white women. If we carried out DNA studies of African American skeletons from a hundred years ago, there is every reason to expect an even greater sex bias.

The genetic patterns suggest that the Thomas Jefferson–Sally Hemings model was replicated countless times by other couples. While this story is one we know about because it is close to us in time and involved famous people, there is every reason to think that sex bias has been central to the history of our species. The genome revolution makes it possible to measure sex bias dating to periods for which we have no records, and thus to begin to understand how inequality may have shaped humanity in deep time.

The Genomic Signature of Inequality

In humans, the profound biological differences that exist between the sexes mean that a single male is physically capable of having far more children than is a single female. Women carry unborn children for nine months and often nurse them for several years prior to having additional children.[13] Men, meanwhile, are able to procreate while investing far less time in the bearing and early rearing of each child, a biological difference whose effects are amplified by social factors such as the fact that in many societies, men are expected to spend little time with their children. So it is that, as measured by the contribution to the next generation, powerful men have the potential to have a far greater impact than powerful women, and we can see this in genetic data.

The great variability among males in the number of offspring produced means that by searching for genomic signatures of past variability in the number of children men have had, we can obtain genetic insights into the degree of social inequality in society as a whole, and not just between males and females. An extraordinary example of this is provided by the inequality in the number of male offspring that seems to have characterized the empire established by Genghis Khan, who ruled lands stretching from China to the Caspian Sea. After his death in 1227, his successors, including several of his sons and grandsons, extended the Mongol Empire even farther—to Korea in the east, to central Europe in the west, and to Tibet in the south. The Mongols maintained rested horses at strategically spaced posts, allowing rapid communication across their more than eight-thousand-kilometer span of territory. The united Mongol Empire was short-lived—for example, the Yüan dynasty they established in China fell in 1368—but their rise to power nevertheless allowed them to leave an extraordinary genetic impact on Eurasia.[14]

A 2003 study led by Christopher Tyler-Smith showed how a relatively small number of powerful males living during the Mongol period succeeded in having an outsize impact on the billions of people living in East Eurasia today.[15] His study of Y chromosomes suggested that one single male who lived around the time of the Mongols left many millions of direct male-line descendants across the territory that the Mongols occupied. The evidence is that about 8 percent of males in the lands that the Mongol Empire once occupied share a characteristic Y-chromosome sequence or one differing from it by just a few mutations. Tyler-Smith and his colleagues called this a "Star Cluster" to reflect the idea of a single ancestor with many descendants, and estimated the date of the founder of this lineage to be thirteen hundred to seven hundred years ago based on the estimated rate of accumulation of mutations on the Y chromosome. The date coincides with that of Genghis Khan, suggesting that this single successful Y chromosome may have been his.

Star Clusters are not limited to Asia. The geneticist Daniel Bradley and his colleagues identified a Y-chromosome type that is present in two to three million people today and derives from an ancestor who lived around fifteen hundred years ago.[16] It is especially common in people with the last name O'Donnell, who descend from one of the most powerful royal families of medieval Ireland, the "Descendants of Niall"—referring to Niall of the Nine Hostages, a legendary warlord from the earliest period of medieval Irish history. If Niall actually existed, he would have lived at about the right time to match the Y-chromosome ancestor.

Star Clusters capture the imagination because they can be tied, albeit speculatively, to historical figures. But the more important point is that Star Cluster analysis provides insights about shifts in social structure that occurred in the deep past that are difficult to get information about in other ways. This is therefore one area in which Y-chromosome and mitochondrial DNA analysis can be instructive, even without whole-genome data. For example, a perennial debate among historians is the extent to which the human past is shaped by single individuals whose actions leave a disproportionate impact on subsequent generations. Star Cluster analysis provides objective information about the importance of extreme inequalities in power at different points in the past.

Two studies, one led by Toomas Kivisild and the other led by Mark Stoneking, have compared the results of Star Cluster analysis on Y-chromosome sequences and on mitochondrial DNA sequences and arrived at an extraordinary result.[17] By counting the number of differences per DNA letter between pairs of sequences, which reflects mutations that accumulated in a clocklike way over time, these studies estimated the time since different pairs of individuals shared common ancestors on the entirely male (Y-chromosome) and entirely female (mitochondrial DNA) lineages.

In mitochondrial DNA data, all the studies found that most couples living in a population today have a very low probability of sharing a common ancestor along their entirely female line in the last ten thousand years, a period postdating the transition to agriculture in many parts of the world. This is exactly as expected if population sizes were large throughout this period. But on the Y chromosome, the studies found a pattern that was strikingly different. In

East Asians, Europeans, Near Easterners, and North Africans, the authors found many Star Clusters with common male ancestors living roughly around five thousand years ago.[18]

The time around five thousand years ago coincides with the period in Eurasia that the archaeologist Andrew Sherratt called the "Secondary Products Revolution," in which people began to find many uses for domesticated animals beyond meat production, including employing them to pull carts and plows and to produce dairy products and clothing such as wool.[19] This was also around the time of the onset of the Bronze Age, a period of greatly increased human mobility and wealth accumulation, facilitated by the domestication of the horse, the invention of the wheel and wheeled vehicles, and the accumulation of rare metals like copper and tin, which are the ingredients of bronze and had to be imported from hundreds or even thousands of kilometers away. The Y-chromosome patterns reveal that this was also a time of greatly increased inequality, a genetic reflection of the unprecedented concentration of power in tiny fractions of the population that began to be possible during this time due to the new economy. Powerful males in this period left an extraordinary impact on the populations that followed them—more than in any previous period—with some bequeathing DNA to more descendants today than Genghis Khan.

From ancient DNA combined with archaeology, we are beginning to build a picture of what this inequality might have meant. The period around five thousand years ago north of the Black and Caspian seas corresponds to the rise of the Yamnaya, who, as discussed in part II, took advantage of horses and wheels to exploit the resources of the open steppe for the first time.[20] The genetic data show that the Yamnaya and their descendants were extraordinarily successful, largely displacing the farmers of northern Europe in the west and the hunter-gatherers of central Asia in the east.[21]

The archaeologist Marija Gimbutas has argued that Yamnaya society was unprecedentedly sex-biased and stratified. The Yamnaya left behind great mounds, about 80 percent of which had male skeletons at the center, often with evidence of violent injuries and buried amidst fearsome metal daggers and axes.[22] Gimbutas argued that the arrival of the Yamnaya in Europe heralded a shift in the power rela-

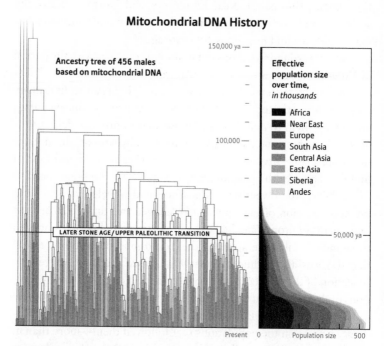

Mitochondrial DNA History

Figure 28a. Human populations have expanded dramatically in the last fifty thousand years. We can see this in trees of relationships constructed based on mitochondrial DNA, where the rarity of recent shared ancestors in this period reflects the large size.

tionships between the sexes. It coincided with the decline of "Old Europe," which according to Gimbutas was a society with little evidence of violence, and in which females played a central social role as is apparent in the ubiquitous Venus figurines. In her reconstruction, "Old Europe" was replaced by a male-centered society, evident not only in the archaeology but also in the male-centered Greek, Norse, and Hindu mythologies of the Indo-European cultures plausibly spread by the Yamnaya.[23]

Any attempt to paint a vivid picture of what a human culture was like before the period of written texts needs to be viewed with caution. Nevertheless, ancient DNA data have provided evidence that the Yamnaya were indeed a society in which power was concentrated

Y Chromosome History

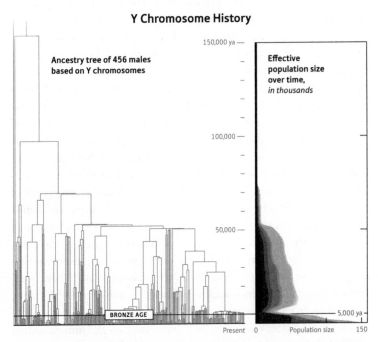

Figure 28b. On the Y chromosome, many people shared ancestors around five thousand years ago. This corresponds to the dawn of the Bronze Age—a period of the first highly socially stratified societies—when some males succeeded in accumulating wealth and making an extraordinary contribution to the next generation.

among a small number of elite males. The Y chromosomes that the Yamnaya carried were nearly all of a few types, which shows that a limited number of males must have been extraordinarily successful in spreading their genes. In contrast, in their mitochondrial DNA, the Yamnaya had more diverse sequences.[24] The descendants of the Yamnaya or their close relatives spread their Y chromosomes into Europe and India, and the demographic impact of this expansion was profound, as the Y-chromosome types they carried were absent in Europe and India before the Bronze Age but are predominant in both places today.[25]

This Yamnaya expansion also cannot have been entirely friendly, as is clear from the fact that the proportion of Y chromosomes of steppe origin in both western Europe[26] and in India[27] today is much larger than the proportion of steppe ancestry in the rest of the genome. This preponderance of male ancestry coming from the steppe implies that male descendants of the Yamnaya with political or social power were more successful at competing for local mates than men from the local groups. The most striking example I know of is from Iberia in far southwestern Europe, where Yamnaya-derived ancestry arrived at the onset of the Bronze Age between forty-five hundred and four thousand years ago. Daniel Bradley's laboratory and my laboratory independently produced ancient DNA from individuals of this period.[28] We found that approximately 30 percent of the Iberian population was replaced along with the arrival of steppe ancestry. However, the replacement of Y chromosomes was much more dramatic: in our data around 90 percent of males who carry Yamnaya ancestry have a Y-chromosome type of steppe origin that was absent in Iberia prior to that time. It is clear that there were extraordinary hierarchies and imbalances in power at work in the expansions from the steppe.

The Star Cluster work rests on Y chromosomes and mitochondrial DNA. What can whole-genome analysis add? When whole-genome data are used to reconstruct the size of the ancestral population of most agricultural groups in the last ten thousand years, they document population growth throughout this period, with no evidence of the Bronze Age population bottlenecks detected from Y chromosomes.[29] This is not what one would expect from averaging the mitochondrial DNA and Y chromosomes. Instead, it is clear that the Y chromosome was a nonrepresentative part of the genome where certain genetic types were more successful at being passed down to later generations than others. In principle, one possible explanation for this is natural selection, whereby some Y chromosomes gave a biological advantage to those who carried them, such as increased fertility. But the fact that this genetic pattern manifested itself around the same time in multiple places around the world—in a period coinciding with the rise of socially stratified societies—is too striking a pattern to be explained by natural selection at multiple independently occurring advantageous mutations. I think a more plausible explanation is that in this period, it began to be possible for single males to accumulate so much power that they could not only gain access to large numbers of females, but they could also pass on their social prestige to subsequent generations and ensure that their male descendants were similarly successful. This process caused the Y chromosomes these males carried to increase in frequency generation after generation, leaving a genetic scar that speaks volumes about past societies.

It is also possible that in this period, individual women began to accumulate more power than they ever had before. Yet because it is biologically impossible for a woman, even a very powerful one, to have an extremely large number of children, the genetic effects of social inequality are much easier to detect on the male line.

Sex Bias in Population Mixture

There are many ways that populations come together—for example through invasions, migrations into each other's homelands, demographic expansion into the same territory, and trade and cultural exchange. Potentially, populations could mix as equals—for example through the overlapping of two equally resourced populations moving peaceably into the same area. But much more often there is asymmetry in the relationship, as reflected in mixture involving males from one group and females from the other, as occurred in the history of African Americans and in the history of the Yamnaya. The different histories of men and women recorded in different parts of the genome make it possible to study this mixture, and thereby to obtain clues about cultural interactions that occurred long ago.

Some of the examples of sex bias evident from genetic data are truly ancient. Take for example the founding of the ancestral population of non-Africans. Any genetic analysis of non-Africans reveals evidence of a population bottleneck dating to some time before fifty thousand years ago—that is, a small number of individuals giving rise to many descendants today. In 2009, I worked with Alon Keinan, a postdoctoral scientist in my laboratory, to compare genetic variation on the X chromosome, the larger of our two sex chromosomes, to the rest of the genome. To our surprise, we found much less genetic variation in non-Africans on chromosome X than would be expected from the level of variation in the rest of the genome, assuming that males and females participated equally in the founding of the ancestral population of non-Africans. The pattern was too extreme to be explained by a simple scenario of more men than women participating in the founding of the ancestral population of non-Africans. But we discovered that one scenario that could explain the pattern is that after the initial budding off of the ancestral population of non-Africans, there was genetic input into this population from males of other groups. Since males carry one copy of chromosome X for every two copies of other chromosomes, a process of repeated waves of male immigration would decrease X-chromosome diversity (meaning that there would be less genetic variation in the population) compared to the rest of the genome, producing the pattern we observed.[30]

This hypothesis gains some plausibility from what we know of the interaction of central African Pygmy hunter-gatherer populations with the Bantu-speaking agriculturalist populations that surround them. When the Bantu first expanded out of west-central Africa several thousand years ago, they had a profound influence on the indigenous rainforest hunter-gatherer populations they encountered, as is evident from the fact that today no Pygmies speak a non-Bantu language and all harbor substantial Bantu-related ancestry. Even today, the overwhelming pattern is that Bantu men mix with Pygmy women and the children are raised in Pygmy communities.[31] The waves of Bantu-related gene flow into the Pygmy population are similar to the scenario that Keinan and I had suggested for the ancestral population of non-Africans. The genetic consequence of this anthropological pattern is reflected in Pygmies having a substantially reduced degree of genetic diversity on chromosome X compared to the expectation from the rest of the genome.[32] Perhaps similar processes were at work in the shared history of non-Africans, explaining the reduced X-chromosome diversity relative to the rest of the genome in that case too.

Evidence of sex bias in the mixture of human populations is

becoming commonplace. The male-biased European contribution to admixed populations in the Americas is stark in African Americans, but it is truly extraordinary in populations in South and Central America, reflecting stories like that of Hernán Cortés and La Malinche. Andrés Ruiz-Linares and colleagues have documented how in the Antioquia region of Colombia, which was relatively isolated between the sixteenth and nineteenth centuries, about 94 percent of the Y chromosomes are European in origin, whereas about 90 percent of the mitochondrial DNA sequences are of Native American origin.[33] This reflects social selection against Native American men. Because nearly all the male ancestry comes from Europeans and nearly all the female ancestry comes from Native Americans, one might naively expect that the people of Antioquia today would derive about half their genome-wide ancestry from Europeans and half from Native Americans, but this is not the case. Instead, about 80 percent of Antioquian ancestry comes from Europeans.[34] The explanation is that Antioquia was flooded by male migrants over many generations. The first European men to arrive mixed with Native American women. Additional European male migrants came later. Through repeated waves of male European migration, the proportion of European ancestry kept increasing everywhere in the genome except for mitochondrial DNA, because mitochondrial DNA is passed to the next generation entirely by females.

Massive sex bias in population mixture also occurred between four thousand and two thousand years ago during the formation of the present-day populations of India.[35] As discussed in part II, endogamous groups in India with traditionally higher social status tend to have more West Eurasian–related ancestry than groups with traditionally lower social status,[36] and the effect is highly sex-biased, as mitochondrial DNA tends to be largely of local origin, whereas a much higher proportion of Y-chromosome types have affinity to West Eurasians.[37] This pattern plausibly reflects a history in which males of West Eurasian–related ancestry were more highly placed in the caste system and sometimes married lower-ranking females. It speaks to a dramatic coming together of socially unequal populations to form the present genetic structure of India.

DNA has the power to overturn expectations from other fields, though, and in this case it has also revealed a surprise about sex-biased mixture. Today, almost every Pacific island population harbors some of its ancestry from people of mainland East Asian origin. As described in part II, this ancestry derives from people whose ancestors originated on Taiwan island and who invented long-distance seafaring and used it to disperse their people, language, and genes. But almost every Pacific island population also harbors Papuan ancestry related to the indigenous hunter-gatherers of the island of New Guinea. Surprisingly, in light of the theme that males from an expanding population tend to mix with local females, initial studies of mitochondrial DNA and Y chromosomes showed that the mixed populations of the Pacific today derive most of their East Asian origin DNA not from male but from female ancestors.[38]

One explanation that has been suggested for this pattern is that in early Pacific island societies, property usually passed down the female line and males were the primary people who moved across islands.[39] But there is another process that may also have contributed. As described in part II, my laboratory showed that the first people of the open Pacific had little Papuan-related ancestry.[40] We showed that later west-to-east waves of migration of people of mixed Papuan and mainland East Asian ancestry explain the ubiquity of Papuan ancestry in the remote Pacific today. If males from this later-arriving population had social advantages relative to the previously resident population, this could have resulted in newly arriving males of primarily Papuan ancestry mixing with previously established females of primarily East Asian–related ancestry.

The Pacific islander example highlights the importance of not simply assuming that genetic analyses of sex-biased events will fulfill expectations from anthropology. Now that the genome revolution has arrived, with its power to reject long-standing theories, we need to abandon the practice of approaching questions about the human past with strong expectations. To understand who we are, we need to approach the past with humility and with an open mind, and to be ready to change our minds out of respect for the power of hard data.

The Future of Genetic Studies of Inequality

At the present time, our methods for using genetic data to study sex bias in human history are frustratingly primitive. Many of the most interesting findings about sex bias so far have been based on just two locations in the genome, the Y chromosome and mitochondrial DNA, which reflect only tiny fractions of our family trees. Studies of sex-biased population dynamics using these sections of the genome become nearly useless for understanding events that occurred more than around ten thousand years ago, because at that time depth, everyone in the world descends from only a small handful of male and female ancestors who are too few in number to support a statistically precise measurement of sex bias.

Future studies of sex-biased mixture, though, will take full advantage of the power of the whole genome. Whole-genome studies can compare the thousands of independent genealogies recorded on chromosome X to the tens of thousands of independent genealogies carried in the rest of the genome. Comparison of genetic variation on the X chromosome to genetic variation elsewhere in the genome should in theory give far more statistical resolution, but while some studies of this sort have been revealing, the accuracy of their estimates has so far been disappointing, which may be because of intense bouts of natural selection that have affected chromosome X more than other chromosomes and that make interpretations of its patterns more difficult. So while many major mixture events—such as those of steppe pastoralists and farmers in Europe, or of Neanderthals and Denisovans with modern humans further back in human prehistory—may well have been sex-biased, detecting them by comparing fractions of ancestry on chromosome X to the rest of the genome is currently challenging.[41] But our present problems with making precise estimates of sex bias based on the X chromosome are to a large extent technical, reflecting the limitations of the statistical techniques currently available. New methods that will be developed over the coming years will unleash the full power of comparison of the X chromosome to the rest of the genome. I hope that these improved methods, along with direct ancient DNA data from people who lived at the times mixture was happening, will enable new genomic insights into the nature of inequality in the deep human past.

The genomic evidence of the antiquity of inequality—between men and women, and between people of the same sex but with greater and lesser power—is sobering in light of the undeniable persistence of inequality today. One possible response might be to conclude that inequality is part of human nature and that we should just accept it. But I think the lesson is just the opposite. Constant effort to struggle against our demons—against the social and behavioral habits that are built into our biology—is one of the ennobling behaviors of which we humans as a species are capable, and which has been critical to many of our triumphs and achievements. Evidence of the antiquity of inequality should motivate us to deal in a more sophisticated way with it today, and to behave a little better in our own time.

11

The Genomics of Race and Identity

Fear of Biological Difference

When I started my first academic job in 2003, I bet my career on the idea that the history of mixture of West Africans and Europeans in the Americas would make it possible to find risk factors that contribute to health disparities for diseases like prostate cancer, which occurs at about a rate 1.7 times higher in African Americans than in European Americans.[1] This particular disparity had not been possible to explain based on dietary and environmental differences across populations, suggesting that genetic factors might play a role.

African Americans today derive about 80 percent of their ancestry from enslaved Africans brought to North America between the sixteenth and nineteenth centuries. In a large group of African Americans, the proportion of African ancestry at any one location in the genome is expected to be close to the average (defining the proportion of African ancestry as the fraction of ancestors that were in West Africa before around five hundred years ago). However, if there are risk factors for prostate cancer that occur at higher frequency in West Africans than in Europeans, then African Americans with prostate cancer are expected to have inherited more African ancestry than the average in the vicinity of these genetic variations. This idea can be used to pinpoint disease genes.

To make such studies possible, I set up a molecular biology laboratory to identify mutations that differed in frequency between West Africans and Europeans. My colleagues and I developed methods that used information from these mutations to identify where in the genome people harbor segments of DNA derived from their West African and European ancestors.[2] To prove that these ideas worked in practice, we applied them to many traits, including prostate cancer, uterine fibroids, late-stage kidney disease, multiple sclerosis, low white blood cell count, and type 2 diabetes.

In 2006, my colleagues and I applied our methods to 1,597 African American men with prostate cancer, and found that in one region of the genome, they had about 2.8 percent more African ancestry than the average in the rest of their genomes.[3] The odds of seeing a rise in African ancestry this large by accident were about ten million to one. When we looked in more detail, we found that this region contained at least seven independent risk factors for prostate cancer, all more common in West Africans than in Europeans.[4] Our findings could account entirely for the higher rate of prostate cancer in African Americans than in European Americans. We could conclude this because African Americans who happen to have entirely European ancestry in this small section of their genomes had about the same risk for prostate cancer as random European Americans.[5]

In 2008, I gave a talk about my work on prostate cancer to a conference on health disparities across ethnic groups in the United States. In my talk, I tried to communicate my excitement about the scientific approach and my conviction that it could help to find genetic risk factors for other diseases. Afterward, though, I was angrily questioned by an anthropologist in the audience, who believed that by studying "West African" or "European" segments of DNA to understand biological differences between groups, I was flirting with racism. Her questions were seconded by several others, and I encountered similar responses at other meetings. A legal ethicist who heard me talk on a similar theme suggested that I might want to refer to the populations from which African Americans descend as "cluster A" and "cluster B." But I replied that it would be dishonest to disguise the model of history that was driving this work. Every feature of the data I looked at suggested that this model was a scientifically meaningful one, providing accurate estimates of where in the genome people harbor segments of DNA from ancestors who lived in West Africa or in Europe in the last twenty generations, prior to the mixture caused by colonialism and the slave trade. It was also clear that the approach was identifying real risk factors for disease that differ in frequency across populations, leading to discoveries with the potential to improve health.

Far from being extremists, my questioners were articulating a mainstream view about the danger of work exploring biological differences among human populations. In 1942, the anthropologist Ashley Montagu wrote *Man's Most Dangerous Myth: The Fallacy of Race*, arguing that race is a social concept and has no biological reality, and setting the tone for how anthropologists and many biologists have discussed this issue ever since.[6] A classic example often cited is the inconsistent definition of "black." In the United States, people tend to be called "black" if they have sub-Saharan African ancestry—even if it is a small fraction and even if their skin color is very light. In Great Britain, "black" tends to mean anyone with sub-Saharan African ancestry who also has dark skin. In Brazil, the definition is different yet again: a person is only "black" if he or she is entirely African in ancestry. If "black" has so many inconsistent definitions, how can there be any biological meaning to "race"?

Beginning in 1972, genetic arguments began to be incorporated into the assertions that anthropologists were making about the lack of substantial biological differences among human populations. In that year, Richard Lewontin published a study of variation in protein types in blood.[7] He grouped the populations he analyzed into seven "races"—West Eurasians, Africans, East Asians, South Asians, Native Americans, Oceanians, and indigenous Australians—and found that around 85 percent of variation in the protein types could be accounted for by variation *within* populations and "races," and only 15 percent by variation *across* them. He concluded: "Races and populations are remarkably similar to each other, with the largest part by far of human variation being accounted for by the differences between individuals. Human racial classification is of no social value and is positively destructive of social and human relations. Since such racial classification is now seen to be of virtually no genetic or taxonomic significance either, no justification can be offered for its continuance."

In this way, through the collaboration of anthropologists and geneticists, a consensus was established that there are no differences among human populations that are large enough to support the concept of "biological race." Lewontin's results made it clear that for the great majority of traits, human populations overlap to such a degree that it is impossible to identify a single biological trait that distinguishes people in any two groups, which is intuitively what some people think of when they conceive of "biological race."

But this consensus view of many anthropologists and geneticists has morphed, seemingly without questioning, into an orthodoxy that the biological differences among human populations are so modest that they should in practice be ignored—and moreover, because the issues are so fraught, that study of biological differences among populations should be avoided if at all possible. It should come as no surprise, then, that some anthropologists and sociologists see genetic research into differences across populations, even if done in a well-intentioned way, as problematic. They are concerned that work on such differences will be used to validate concepts of race that should be considered discredited. They see this work as located on a slippery slope to the kinds of pseudoscientific arguments about biological difference that were used in the past to try to justify the slave trade, the eugenics movement to sterilize the disabled as biologically defective, and the Nazis' murder of six million Jews.

The concern is so acute that the political scientist Jacqueline Stevens has even suggested that research and even emails discussing biological differences across populations should be banned, and that the United States "should issue a regulation prohibiting its staff or grantees . . . from publishing in any form—including internal documents and citations to other studies—claims about genetics associated with variables of race, ethnicity, nationality, or any other category of population that is observed or imagined as heritable unless statistically significant disparities between groups exist and description of these will yield clear benefits for public health, as deemed by a standing committee to which these claims must be submitted and authorized."[8]

The Language of Ancestry

But whether we like it or not, there is no stopping the genome revolution. The results that it is producing are making it impossible to maintain the orthodoxy established over the last half century, as they are revealing hard evidence of substantial differences across populations.

The first major engagement between the genome revolution and anthropological orthodoxy came in 2002, when Marc Feldman and his colleagues showed that by studying enough places in the genome—they analyzed 377 variable positions—it is possible to group most people in a worldwide population sample into clusters that correlate strongly to popular categories of race in the United States: "African," "European," "East Asian," "Oceanian," or "Native American."[9] While Feldman's conclusions were broadly consistent with Lewontin's in that his data also showed more variation within groups than among them, his study defined clusters in terms of combinations of mutations instead of looking at mutations individually as Lewontin had done.

Scientists were quick to respond. One was Svante Pääbo, who eight years later would go on to lead the work to sequence whole genomes of archaic Neanderthals and Denisovans. Pääbo came to the debate about the nature of human population structure as a founding director of the Max Planck Institute for Evolutionary Anthropology in Leipzig, which was set up in 1997 in an effort to return Germany to a field in which it had played a leading role before the Second World War but that it had largely abandoned due to anthropologists' central contribution to developing Nazi race theory.

Pääbo took seriously his moral responsibility as head of an ambitious German institute of anthropology, and wondered whether the truth about human population structure could be more like the anthropologist Frank Livingston's suggestion that "there are no races, there are only clines"—a view in which human genetic variation is characterized by gradual geographic gradients that reflect interbreeding among neighbors.[10] To explore this possibility, Pääbo investigated whether the clusters the Feldman study found appeared sharply defined because the analyzed populations had been sampled in a nonrandom fashion across the world. To understand how nonrandom sampling could contribute to this result, consider the United States, which harbors extraordinary diversity, but where genetic discontinuities among groups such as African Americans, European Americans, and East Asians are sharper than in the places from which immigrant populations came because the United States has drawn its immigrants from a subset of world locations. For example, in the United States, most of the African ancestry is from a handful of groups in West Africa,[11] most of the European ancestry is from northwest Europe, and most of the Asian ancestry is from Northeast Asia. Pääbo showed that such nonrandom sampling could account for some of the effects Feldman and colleagues observed. However, later work proved that nonrandom sampling could not account for most of the structure, as substantial clustering of human populations is observed even when repeating analyses on geographically more evenly distributed sets of samples.[12]

Another flurry of discussion followed a 2003 paper led by Neil Risch, who argued that racial grouping is useful in medical research, not just to adjust for socioeconomic and cultural differences, but also because it correlates with genetic differences that are important to know about when diagnosing and treating disease.[13] Risch was convinced by examples like sickle cell disease, which occurs far more often in African Americans than in other populations in the United States. He argued that it was appropriate for doctors to be more likely to think of sickle cell disease if the patient is African American.

In 2005, the U.S. Food and Drug Administration lent support to this way of thinking when it approved BiDil, a combination of two medications approved to treat heart failure in African Americans because data suggested it was more effective in African Americans than in European Americans. But on the other side of the argument, David Goldstein suggested that U.S. racial categories are so weakly predictive of most biological outcomes that they do not have long-term value.[14] He and his colleagues showed that the frequencies of genetic variants that determine dangerous reactions to drugs are poorly predicted by U.S. census categories. He acknowledged that the reliance on racial and ethnic categories is useful given our poor present knowledge, but predicted that the future will involve testing individuals directly for what mutations they have, and doing away altogether with racial classification as a basis for making individualized decisions about care.

Against this backdrop of controversy emerged work like mine, focusing on methods to determine population origin not just of our ancestors but also of individual segments of our genomes. The anthropologist Duana Fullwiley has written that the development of what she calls "admixture technology" and the language of "ancestry" that geneticists like me have adopted is a reversion to traditional ideas of biological race.[15] She has pointed out that in the United States, the "ancestry" terms that we use map relatively closely to traditional racial categories, and her view is that the population genetics community has invented a set of euphemisms to discuss topics that had become taboo. The belief that we have embraced euphemisms is also shared by some on the other side of the political spectrum. At a 2010 meeting I attended at Cold Spring Harbor Laboratory, the journalist Nicholas Wade described his resentment of the population genetics community's "ancestry" terminology, asserting that "race is a perfectly good English word."

But "ancestry" is not a euphemism, nor is it synonymous with "race." Instead, the term is born of an urgent need to come up with a precise language to discuss genetic differences among people at a time when scientific developments have finally provided the tools to detect them. It is now undeniable that there are nontrivial average genetic differences across populations in multiple traits, and the race vocabulary is too ill-defined and too loaded with historical baggage to be helpful. If we continue to use it we will not be able to escape the current debate, which is mired in an argument between two indefensible positions. On the one side there are beliefs about the nature of the differences that are grounded in bigotry and have little basis in reality. On the other side there is the idea that any biological differences among populations are so modest that as a matter of social policy they can be ignored and papered over. It is time to move on from this paralyzing false dichotomy and to figure out what the genome is actually telling us.

Real Biological Difference

I have deep sympathy for the concern that genetic discoveries about differences among populations may be misused to justify racism. But it is precisely because of this sympathy that I am worried that people who deny the possibility of substantial biological differences among populations across a range of traits are digging themselves into an indefensible position, one that will not survive the onslaught of science. In the last couple of decades, most population geneticists have sought to avoid contradicting the orthodoxy. When asked about the possibility of biological differences among human populations, we have tended to obfuscate, making mathematical statements in the spirit of Richard Lewontin about the average difference between individuals from within any one population being around six times greater than the average difference between populations. We point out that the mutations that underlie some traits that differ dramatically across populations—the classic example is skin color—are unusual, and that when we look across the genome it is clear that the typical differences in frequencies of mutations across populations are far less.[16] But this carefully worded formulation is deliberately masking the possibility of substantial average differences in biological traits across populations.

To understand why it is no longer an option for geneticists to lock arms with anthropologists and imply that any differences among human populations are so modest that they can be ignored, go no further than the "genome bloggers." Since the genome revolution began, the Internet has been alive with discussion of the papers written about human variation, and some genome bloggers have even become skilled analysts of publicly available data. Compared to most academics, the politics of genome bloggers tend to the right—Razib Khan[17] and Dienekes Pontikos[18] post on findings of average differences across populations in traits including physical appearance and athletic ability. The *Eurogenes* blog spills over with sometimes as many as one thousand comments in response to postings on the charged topic of which ancient peoples spread Indo-European languages,[19] a highly sensitive issue since as discussed in part II, narratives about the expansion of Indo-European speakers have been used as a basis for building national myths,[20] and sometimes have been abused as happened in Nazi Germany.[21] The genome bloggers' political beliefs are fueled partly by the view that when it comes to discussion about biological differences across populations, the academics are not honoring the spirit of scientific truth-seeking. The genome bloggers take pleasure in pointing out contradictions between the politically correct messages academics often give about the indistinguishability of traits across populations and their papers showing that this is not the way the science is heading.

What real differences do we know about? We cannot deny the existence of substantial average genetic differences across populations, not just in traits such as skin color, but also in bodily dimensions, the ability to efficiently digest starch or milk sugar, the ability to breathe easily at high altitudes, and susceptibility to particular diseases. These differences are just the beginning. I expect that the reason we don't know about a much larger number of differences among human populations is that studies with adequate statistical power to detect them have not yet been carried out. For the great majority of traits, there is, as Lewontin said, much more variation within populations than across populations. This means that individuals with extreme high or low values of the great majority of traits can occur in any population. But it does not preclude the existence of subtler, average differences in traits across populations.

The indefensibility of the orthodoxy is obvious at almost every turn. In 2016, I attended a lecture on race and genetics by the biologist Joseph L. Graves Jr. at the Peabody Museum of Archaeology and Ethnography at Harvard. At one point, Graves compared the approximately five mutations known to have large effects on skin pigmentation and that are obviously different in frequency across populations to the more than ten thousand genes known to be active in human brains. He argued that in contrast to pigmentation genes, the patterns at genes particularly active in the brain would surely average out over so many locations, with some mutations nudging cognitive and behavioral traits in one direction and some pushing in the other direction. But this argument doesn't work, because in fact, if natural selection has exerted different pressures on two populations since they separated, traits influenced by many mutations are just as capable of achieving large average differences across populations as traits influenced by few mutations. And indeed, it is already known that traits shaped by many mutations (as is probably the case for behavior and cognition) are at least as important targets of natural selection as traits like skin color that are driven by a small number of mutations.[22] The best example we currently have of a trait governed by many mutations is height. Studies in hundreds of thousands of people have shown that height is determined by thousands of variable positions across the genome. A 2012 analysis led by Joel Hirschhorn showed that natural selection on these is responsible for the shorter average height in southern Europeans compared to northern Europeans.[23] Height isn't the only example. Jonathan Pritchard led a study showing that in the last approximately two thousand years there has been selection for genetic variations that affect many other traits in Britain, including an increase in average infant head size and an increase in average female hip size (possibly to accommodate the increased higher average infant head size during childbirth).[24]

It is tempting to argue that genetic influence on bodily dimensions is one thing, but that cognitive and behavioral traits are another. But this line has already been crossed. Often when a person participates in a genetic study of a disease, he or she fills out a form providing information on height, weight, and number of years of education. By compiling the information on the number of years of education for over four hundred thousand people of European ancestry whose genomes have been surveyed in the course of various disease studies, Daniel Benjamin and colleagues identified seventy-four genetic variations each of which has overwhelming evidence of being more common in people with more years of education than in people with fewer years even after controlling for such possibly confounding factors as heterogeneity in the study population.[25] Benjamin and colleagues also showed that the power of genetics to predict number of years of education is far from trivial, even though social influences surely have a greater average influence on this behavior than genetics. They showed that in the European ancestry population in which they carried out their study, it should be possible to build a genetic predictor in which the probability of completing twelve years of education is 96 percent for the twentieth of people with the highest prediction compared to 37 percent for the lowest.[26]

How do these genetic variations influence educational attainment? The obvious guess is that they have a direct effect on academic abilities, but that is probably wrong. A study of more than one hundred thousand Icelanders showed that the variations also increase the age at which a woman has her first child, and that this is a more powerful effect than the one on the number of years of education. It is possible that these variations exert their effect indirectly, by nudging people to defer having children, which makes it easier for them to complete their education.[27] This shows that when we discover biological differences governing behavior, they may not be working in the way we naively assume.

Average differences across populations in the frequencies of the mutations that affect educational attainment have not yet been identified. But a sobering finding is that older people in Iceland are systematically different from younger people in having a higher genetically predicted number of years of education.[28] Augustine Kong, the lead author of the Icelandic study, showed that this reflects natural selection over the last century against people with more predicted education, likely because of selection for people who began having children at a younger age. Given that the genetic underpinnings of the number of years of education a person achieves have measurably changed within a century in a single population under the pressure of natural selection, it seems highly likely that the trait differs across populations too.

No one knows how the genetic variations that influence educational attainment in people of European ancestry affect behavior in people of non-European ancestries, or in differently structured social systems. That said, it seems likely that if these mutations have an effect on behavior in one population they will do so in others, too, even if the effects differ by social context. And educational attainment as a trait is likely to be only the tip of an iceberg of behavioral traits affected by genetics. The Benjamin study has already been joined by others finding genetic predictors of behavioral traits,[29] including one of more than seventy thousand people that found mutations in more than twenty genes that were significantly predictive of performance on intelligence tests.[30]

For those who wish to argue against the possibility of biological differences across populations that are substantial enough to make a difference in people's abilities or propensities, the most natural refuge might be to make the case that even if such differences exist, they will be small. The argument would be that even if there are average differences across human populations in genetically determined traits affecting cognition or behavior, so little time has passed since the separation of populations that the quantitative differences across populations are likely to be trivially small, harkening back to Lewontin's argument that the average genetic difference between populations is much less than the average difference between individuals. But this argument doesn't hold up either. The average time separation between pairs of human populations since they diverged from common ancestral populations, which is up to around fifty thousand years for some pairs of non-African populations, and up to two hundred thousand years or more for some pairs of sub-Saharan African populations, is far from negligible on the time scale of human evolution. If selection on height and infant head circumference can occur within a couple of thousand years,[31] it seems a bad bet to argue that there cannot be similar average differences in cognitive or behavioral traits. Even if we do not yet know what the differences are, we should prepare our science and our society to be able to deal with the reality of differences instead of sticking our heads in the sand and pretending that differences cannot be discovered. The approach of staying mum, of implying to the public and to colleagues that substantial differences in traits across populations are unlikely to exist, is a strategy that we scientists can no longer afford, and that in fact is positively harmful. If as scientists we willfully abstain from laying out a rational framework for discussing human differences, we will leave a vacuum that will be filled by pseudoscience, an outcome that is far worse than anything we could achieve by talking openly.

The Genome Revolution's Insight

On the question of whether traditional social categories of race correspond to meaningful biological categories, the genome revolution has already provided us with new insights that go far beyond the information that was available to the first population geneticists and anthropologists who grappled with the issue. In this way, the data provided by the genome revolution are potentially liberating, providing an opportunity for intellectual progress beyond the current stale framing of the debate.

As recently as 2012, it still seemed reasonable to interpret human genetic data as pointing to immutable categories such as "East Asians," "Caucasians," "West Africans," "Native Americans," and "Australasians," with each group having been separated and unmixed for tens of thousands of years. The 2002 study led by Marc Feldman produced clusters that corresponded relatively well to these categories, and the model seemed to be doing a good job of describing variation in many parts of the world (with some exceptions).[32] In other papers, Feldman and his colleagues proposed a model for how this kind of structure could arise among human populations. Their proposal was that modern humans expanding out of Africa and the Near East after around fifty thousand years ago left descendant populations along the way, which in turn budded off their own descendant populations, with the present-day inhabitants of each region being descended directly from the modern humans who first arrived.[33] Their "serial founder" model was more sophisticated than that imagined by biological race theorists in the seventeenth to twentieth centuries, but shared with it the prediction that after being established, human populations hardly mixed with each other.

But ancient DNA discoveries have rendered the serial founder model untenable. We now know that the present-day structure of populations does not reflect the one that existed many thousands of years ago.[34] Instead, the current populations of the world are mixtures of highly divergent populations that no longer exist in unmixed form—for example, the Ancient North Eurasians, who contributed a large amount of the ancestry of present-day Europeans as well as of Native Americans,[35] and multiple ancient populations of the Near East, each as differentiated from the other as Europeans and East Asians are differentiated from each other today.[36] Most of today's populations are not exclusive descendants of the populations that lived in the same locations ten thousand years ago.

The findings that the nature of human population structure is not what we assumed should serve as a warning to those who think they know that the true nature of human population differences will correspond to racial stereotypes. Just as we had an inaccurate picture of early human origins before the ancient DNA revolution unleashed an avalanche of surprises, so we should distrust the instincts that we have about biological differences. We do not yet have sufficient sample sizes to carry out compelling studies of most cognitive and behavioral traits, but the technology is now available, and once high-quality studies are performed—which they will be somewhere in the world whether we like it or not—any genetic associations they find will be undeniable. We will need to deal with these studies and react responsibly to them when they are published, but we can already be sure that we will be surprised by some of the outcomes.

Unfortunately, today there is a new breed of writers and scholars who argue not only that there are average genetic differences, but that they can guess what they are based on traditional racial stereotypes.

The person who has most recently made a prominent argument that there is a genetic basis to stereotypes about differences across human populations is the *New York Times* journalist Nicholas Wade, who in 2014 published *A Troublesome Inheritance: Genes, Race and Human History*.[37] The abiding theme of Wade's reporting is the propensity of academics to band together to enforce orthodoxies and to be shown up by a band of rebels speaking the truth (he has written on scientific fraud, described the Human Genome Project as a monolith wastefully spending the public's money, and attacked the value of genome-wide association studies for finding common genetic variations contributing to risk for diseases). Wade's *Troublesome Inheritance* ran with the theme again, suggesting that a politically correct alliance of anthropologists and geneticists has banded together to suppress the truth that there are significant differences among human populations and that those differences correspond to classic stereotypes. One part of the argument has something to it—Wade correctly highlights the problem of an academic community trying to enforce an implausible orthodoxy. Yet the "truth" that he puts forward in opposition, the idea that not only are there substantial differences, but that they likely correspond to traditional racial stereotypes, has no merit. Wade's book combines compelling content with parts that are entirely speculative, presenting everything with the same authority and in the same voice, so that naive readers who accept the parts of it that are well argued are tempted to accept the rest. Worse, when compared to Wade's previous writing, in which the rebels speaking the truth were scholars of creativity and accomplishment, he does not identify any serious scholarship in genetics supporting his speculations.[38] And yet by celebrating those who have opposed the flawed orthodoxy, he implies wrongly that their alternative theories must be right.

As an example of the speculations to which Wade gives pride of place, one of his chapters focuses on a 2006 essay by Gregory Cochran, Jason Hardy, and Henry Harpending suggesting that the high average intelligence quotient (IQ) of Ashkenazi Jews (more than one standard deviation above the world average), and their disproportionate share of Nobel Prizes (about one hundred times the world average), might reflect natural selection due to a millennium-long history in which Jewish populations practiced moneylending, a profession that required writing and calculation.[39] They also pointed to the high rate in Ashkenazi Jews of Tay-Sachs disease and Gaucher disease, which are due to mutations that affect storage of fat in brain cells, and which they hypothesized rose in frequency under the pressure of selection for genetic variations contributing to intelligence (they argued that these mutations might be beneficial when they occur in one copy rather than the two needed to cause disease). This argument is contradicted by the evidence that these diseases almost certainly owe their origin to random bad luck—the fact that during the medieval population bottleneck that affected Ashkenazi Jews, the small number of individuals who had many descendants happened to carry these mutations[40]—yet Wade highlights the work on the basis that it *might* be right. Harpending has a track record of speculating without evidence on the causes of behavioral differences among populations. In a talk he gave at a 2009 conference on "Preserving Western Civilization," he asserted that people of sub-Saharan African ancestry have no propensity to work when they don't have to— "I've never seen anyone with a hobby in Africa," he said—because, he thought, sub-Saharan Africans have not gone through the type of natural selection for hard work in the last thousands of years that some Eurasians had.[41]

Wade also highlighted *A Farewell to Alms*, a book by the economist Gregory Clark suggesting that the reason the Industrial Revolution took off in Britain before it did elsewhere was the relatively high birth rate among wealthy people in Britain for the preceding five centuries compared to less wealthy people. Clark argued that this higher birth rate spread through the population the traits needed for a capitalist surge, including individualism, patience, and willingness to work long hours.[42] Clark admits that he cannot distinguish between the transmission of genes and the transmission of culture across the generations, but Wade nevertheless takes his argument as evidence that genetics *might* have played a role.

I have spent some space discussing errors in Wade's book because I feel it is important to explain that just because many academics have been engaged in trying to maintain an implausible orthodoxy, it does not mean that every unorthodox "heretic" is right. And yet Wade suggests precisely this. He writes, "Each of the major civilizations has developed the institutions appropriate for its circumstances and survival. But these institutions, though heavily imbued with cultural traditions, rest on a bedrock of genetically shaped human behavior. And when a civilization produces a distinctive set of institutions that endures for many generations, that is the sign of a supporting suite of variations in the genes that influence human social behavior."[43] In a written version of a nod and a wink, Wade is suggesting that popular racist ideas about the differences that exist among populations have something to them.

Wade is far from the only person who is convinced he knows the truth about the differences among populations. At the same 2010 meeting on "DNA, Genetics, and the History of Mankind" at which

I first met Wade, I heard a rustling behind my shoulder and turned with a shock to see James Watson, who in 1953 codiscovered the structure of DNA. Watson had until a few years earlier been the director of the Cold Spring Harbor Laboratory at which the meeting was held. A century ago, the laboratory was the epicenter of the eugenics movement in the United States, keeping records on traits in many people to help guide selective breeding, and lobbying for legislation that was passed in many states to sterilize people considered to be defective and to combat a perceived degradation of the gene pool. It was ironic, then, that Watson was forced to retire as head of Cold Spring Harbor after being quoted in an interview with the British *Sunday Times* newspaper as having said that he was "inherently gloomy about the prospect of Africa," adding that "[all] our social policies are based on the fact that their intelligence is the same as ours—whereas all the testing says not really."[44] (No genetic evidence for this claim exists.) When I saw Watson at Cold Spring Harbor, he leaned over and whispered to me and to the geneticist Beth Shapiro, who was sitting next to me, something to the effect of "When are you guys going to figure out why it is that you Jews are so much smarter than everyone else?" He then said that Jews and Indian Brahmins were both high achievers because of genetic advantages conferred by thousands of years of natural selection to be scholars. He went on to whisper that Indians in his experience were also servile, much like he thought they had been under British colonialism, and he speculated that this trait had come about because of selection under the caste system. He also talked about how East Asian students tended to be conformist, because of selection for conformity in ancient Chinese society.

The pleasure Watson takes in challenging establishment views is legendary. His obstreperousness may have been important to his success as a scientist. But now as an eighty-two-year-old man, his intellectual rigor was gone, and what remained was a willingness to vent his gut impressions without subjecting them to any of the testing that characterized his scientific work on DNA.

Writing now, I shudder to think of Watson, or of Wade, or their forebears, behind my shoulder. The history of science has revealed, again and again, the danger of trusting one's instincts or of being led astray by one's biases—of being too convinced that one knows the truth. From the errors of thinking that the sun revolves around the earth, that the human lineage separated from the great ape lineage tens of millions of years ago, and that the present-day human population structure is fifty thousand years old whereas in fact we know that it was forged through population mixtures largely over the last five thousand years—from all of these errors and more, we should take the cautionary lesson not to trust our gut instincts or the stereotyped expectations we find around us. If we can be confident of anything, it is that whatever differences we think we perceive, our expectations are most likely wrong. What makes Watson's and Wade's and Harpending's statements racist is the way they jump from the observation that the academic community is denying the possibility of differences that are plausible, to a claim with no scientific evidence[45] that they know what those differences are and also that the differences correspond to long-standing popular stereotypes—a conviction that is essentially guaranteed to be wrong.

We truly have no idea right now what the nature or direction of genetically encoded differences among populations will be. An example is the extreme overrepresentation of people of West African ancestry among elite sprinters. All the male finalists in the Olympic hundred-meter race since 1980, even those from Europe and the Americas, had recent West African ancestry.[46] The genetic hypothesis most often invoked to explain this is that there has been an upward shift in the average sprinting ability of people of West African ancestry due to natural selection. A small increase in the average might not sound like much, but it can make a big difference at the extremes of high ability—for example, a 0.8-standard-deviation increase in the average sprinting ability in West Africans would be expected to lead to a hundredfold enrichment in the proportion of people above the 99.9999999th percentile point in Europeans. But an alternative explanation that would predict the same magnitude of effect is that there is simply more variation in sprinting ability in people of West African ancestry—with more people of both very high and very low abilities.[47] A wider spread of abilities around the same mean and a hundredfold enrichment in West Africans in the proportion of people above the 99.9999999th percentile point seen in Europeans is in fact exactly what is expected given the approximately 33 percent higher genetic diversity in West Africans than in Europeans.[48] Whether or not this explains the dominance of West Africans in sprinting, for many biological traits—including cognitive ones—there is expected to be a higher proportion of sub-Saharan Africans with extreme genetically predicted abilities.

So how should we prepare for the likelihood that in the coming years, genetic studies will show that behavioral or cognitive traits are influenced by genetic variation, and that these traits will differ on average across human populations, both with regard to their average and their variation within populations? Even if we do not yet know what those differences will be, we need to come up with a new way of thinking that can accommodate such differences, rather than deny categorically that differences can exist and so find ourselves caught without a strategy once they are found.

It would be tempting, in the wake of the genome revolution, to settle on a new comforting platitude, invoking the history of repeated admixture in the human past as an argument for population differences being meaningless. But such a statement is wrongheaded, as if we were to randomly pick two people living in the world today, we would find that many of the population lineages contributing to them have been isolated from each other for long enough that there has been ample opportunity for substantial average biological differences to arise between them. The right way to deal with the inevitable discovery of substantial differences across populations is to realize that their existence should not affect the way we conduct ourselves. As a society we should commit to according everyone equal rights despite the differences that exist among individuals. If we aspire to treat all individuals with respect regardless of the extraordinary differences that exist among individuals within a population, it should not be so much more of an effort to accommodate the smaller but still significant average differences across populations.

Beyond the imperative to give everyone equal respect, it is also important to keep in mind that there is a great diversity of human traits, including not just cognitive and behavioral traits, but also areas of athletic ability, skill with one's hands, and capacity for social interaction and empathy. For most traits, the degree of variation among individuals is so large that any one person in any population can excel at any trait regardless of his or her population origin, even if particular populations have different average values due to a mixture of genetic and cultural influences. For most traits, hard work and the right environment are sufficient to allow someone with a lower genetically predicted performance at some task to excel compared to people with a higher genetically predicted performance. Because of the multidimensionality of human traits, the great variation that exists among individuals, and the extent to which hard work and upbringing can compensate for genetic endowment, the only sensible approach is to celebrate every person and every population as an extraordinary realization of our human genius and to give each person every chance to succeed, regardless of the particular average combination of genetic propensities he or she happens to display.

For me, the natural response to the challenge is to learn from the example of the biological differences that exist between males and females. The differences between the sexes are in fact more profound than those that exist among human populations, reflecting more than a hundred million years of evolution and adaptation. Males and females differ by huge tracts of genetic material—a Y chromosome that males have and that females don't, and a second X chromosome that females have and males don't. Most people accept that the biological differences between males and females are profound, and that they contribute to average differences in size and physical strength as well as in temperament and behavior, even if there are questions about the extent to which particular differences are also influenced by social expectations and upbringing (for example, many of the jobs in industry and the professions that women fill in great numbers today had few women in them a century ago). Today we aspire both to recognize that biological differences exist and to accord everyone the same freedoms and opportunities regardless of them. It is clear from the abiding average inequities that persist between women and men that fulfilling these aspirations is a challenge, and yet it is important to accommodate and even embrace the real differences that exist, while at the same time struggling to get to a better place.

The real offense of racism, in the end, is to judge individuals by a supposed stereotype of their group—to ignore the fact that when applied to specific individuals, stereotypes are almost always mis-leading. Statements such as "You are black, you must be musical" or "You are Jewish, you must be smart" are unquestionably very harmful. Everyone is his or her own person with unique strengths and weaknesses, and should be treated as such. Suppose you are the coach of a track-and-field team, and a young person walks on and asks to try out for the hundred-meter race, in which people of West African ancestry are statistically highly overrepresented, suggesting the possibility that genetics may play a role. For a good coach, race is irrelevant. Testing the young person's sprinting speed is simple—take him or her out to the track to run against the stopwatch. Most situations are like this.

A New Basis for Identity

The genome revolution is actually a far more effective force for com-ing to a new understanding of human difference and identity—for understanding our own personal place in the world around us—than for promoting old beliefs that more often than not are mistaken.

To understand the power of the genome revolution for under-mining old stereotypes about identity and building up a new basis for identity, consider how its finding of repeated mixture in human history has destroyed nearly every argument that used to be made for biologically based nationalism. The Nazi ideology of a "pure" Indo-European-speaking Aryan race with deep roots in Germany, traceable through artifacts of the Corded Ware culture, has been shattered by the finding that the people who used these artifacts came from a mass migration from the Russian steppe, a place that German nationalists would have despised as a source.[49] The Hindu-tva ideology that there was no major contribution to Indian culture from migrants from outside South Asia is undermined by the fact that approximately half of the ancestry of Indians today is derived from multiple waves of mass migration from Iran and the Eurasian steppe within the last five thousand years.[50] Similarly, the idea that the Tutsis in Rwanda and Burundi have ancestry from West Eurasian farmers that Hutus do not—an idea that has been incorporated into arguments for genocide[51]—is nonsense. We now know that nearly every group living today is the product of repeated population mixtures that have occurred over thousands and tens of thousands of years. Mixing is in human nature, and no one population is—or could be—"pure."

Nonscientists have already realized the potential of the genome revolution for forming new narratives. African Americans have been at the forefront of this movement. During the slave trade, Africans were uprooted and forcibly deprived of their culture, with the effect that within a few generations much of their ancestors' religion, language, and traditions were gone. In 1976, Alex Haley's novel *Roots* used literature to begin to reclaim lost roots by recounting the odyssey of the slave Kunta Kinte and his descendants.[52] Following in this tradition, Harvard professor of literature Henry Louis Gates Jr. has capitalized on the potential of genetic studies to recover lost roots for African Americans. In his *Faces of Americans* television series and the *Finding Your Roots* series that followed it, he declares to the cellist Yo-Yo Ma, who is able to trace his ancestry back to thirteenth-century China, that Gates, as an African American, will never know how that feels, but he shows that genetics can provide richly informative insights even for African Americans with limited genealogical records.[53]

A new industry, "personal ancestry testing," has sprung up to capitalize on the potential of the genome revolution to form the basis for new narratives and to compare the genomes of consumers to others who have already been tested. The television programs that Gates has produced have been built around the idea of tracing the genealogies and DNA of celebrity guests, using the literary device of telling the personal stories of famous people to help viewers understand the power of genetic data to reveal features of their family's past about which they could not otherwise have been aware. For example, the programs revealed unknown deep relationships between pairs of guests on the program (shared ancestors within the last few hundred years). They also used genetic tests to determine not only the continents on which people's ancestors lived, but also the regions within continents.

As a white person in the United States with its history of forcible deprivation of peoples of their roots, I feel that everyone—African Americans and Native Americans especially—has the right to try to use genetic data to help fill in missing pieces in his or her family history. Nevertheless, for those who assume that personal ancestry testing results have the authority of science, it is important to keep in mind that many of the results are easily misinterpreted and rarely include the warnings that scientists attach to tentative findings.

Some of the best examples come from the industry that sprang up to provide genetic results to African Americans. One company is African Ancestry, which provides customers with information on the West African tribe and country in which their Y-chromosome or mitochondrial DNA type is most common. Such results are easy to overinterpret, as the frequencies of Y-chromosome and mitochondrial DNA types are too similar across West Africa to make exact determinations with confidence. As an example, consider a Y-chromosome type that is carried slightly more often in the Hausa ethnic group than in the neighboring Yoruba, Mende, Fulani, and Beni groups. When African Ancestry sends its report, it might state that an African American man has a Y-chromosome type that is most common in the Hausa.[54] But it is quite possible and even likely that the true ancestor was not the Hausa, because there are many tribes in West Africa, and no one tribe contributed more than a modest fraction of the African ancestry of African Americans.[55] And yet people who have taken these tests often return with the impression that they know their origin. The geneticist Rick Kittles, a population geneticist who is the cofounder of African Ancestry, described this feeling, asserting, "My female line goes back to northern Nigeria, the land of the Hausa tribe. I then went to Nigeria and talked to people and learned about the Hausa's culture and tradition. That gave me a sense about who I am."[56] Whole-genome ancestry tests in theory have much more power than tests based on Y chromosomes and mitochondrial DNA. But at present, even whole-genome methods are not good enough to provide high-resolution information about where the ancestors of an African American person lived within Africa, in part because the databases of present-day populations in West Africa are not complete enough. Much more research needs to be done to make it possible to carry out studies like these with any reliability.

For African Americans, another frustration may be that the cultural upheaval that occurred after African slaves arrived in North America has been so enormous that today there are few differences among African Americans with respect to the places in Africa from which their ancestors came. Africans from one part of the continent were traded around and mixed with those from another, with the result that within a few generations the great cultural diversity and variation of ancestry that existed among the first slaves were blurred to the point of unrecognizability. The nearly complete homogenization of African ancestry that occurred was evident in an unpublished study I carried out in 2012 with Kasia Bryc, who analyzed genome-wide data from more than fifteen thousand African Americans from Chicago, New York, San Francisco, Mississippi, North Carolina, and the South Carolina Sea Islands, and tested if some African American populations were more closely related to particular West Africans than others, as might be expected based on the heterogeneous supply routes for U.S. slaves.[57] It made sense to expect some differences. Of the four big slave ports, New Orleans was supplied mostly by French slave traders, whereas Baltimore, Savannah, and Charleston were supplied mostly by the British drawing from different points in Africa. But what we found is that the mixing of the West African ancestors of African Americans has been so thorough that we could not detect any differences in the African source populations for mainland populations. Only in the Sea Islands off South Carolina did we detect evidence of a particular connection to one place in Africa, in this case to people of the country of Sierra Leone, the place of origin of the language with an African grammar still spoken by Gullah Sea Islanders. It will take ancient DNA studies of first-generation enslaved Africans to actually trace roots to Africa.[58]

The problem with the results sometimes provided by personal ancestry testing companies is not limited to African Americans. It is a more general pitfall that stems from the financial incentive that such companies have to provide people with what feel like meaningful findings. This is a problem even for the most rigorous of the companies. Between 2011 and 2015, the genetic testing company 23andMe provided customers with an estimate of their proportion of Neanderthal ancestry, allowing them to make a personal connection to the research showing that non-Africans derive around 2 percent of their genomes from Neanderthals.[59] The measurement made by the test was highly inaccurate, however, since the true variation in Neanderthal proportion within most populations is only a few tenths of a percent, and the test reports variation of a few percentage points.[60] Several people have told me excitedly that their 23andMe Neanderthal testing result put them in the top few percent of people in the world in Neanderthal ancestry, but because of the test's inaccuracies, the probability that people who got such a high 23andMe Neanderthal reading really do have more than the average proportion of Neanderthal ancestry is only slightly greater than 50/50. I raised this problem to members of the 23andMe team and even highlighted the problems in a 2014 scientific paper.[61] Later, 23andMe changed its report to no longer provide these statements. However, the company continues to provide its customers with a ranking of the number of Neanderthal-derived mutations they carry.[62] This ranking, too, does not provide strong evidence that customers have inherited more Neanderthal DNA than their population average.

Not all the findings reported by the personal ancestry companies are inaccurate, and many people have obtained what for them is satisfying information from such testing, especially when it comes to tracing genealogies where the paper trail runs cold. One example is adoptees seeking their biological parents. Another is tracking down extended families.

From my own perspective, though, I do not find this approach to be satisfying. In preparing to write this book, I considered whether I should send my DNA to a personal testing company or study it in my own lab, and then describe the results, in imitation of the approach taken by many journalists covering the field of personal ancestry testing. But honestly, I am not interested. My own group—Ashkenazi Jews—is already overstudied. I am confident that my genome will be much like that of anyone else from this population. I would much rather use any resources I have to sequence the genomes of people who are understudied. I am also worried about the intellectual pitfall of self-study. I am innately suspicious of scientists who are hyper-interested in their own family or culture. They simply care too much. In my own laboratory, there are researchers from all over the world, and I encourage them, not always successfully, to choose projects on peoples not their own. For me, the approach of using the genome as a tool to connect myself to the world around me through personal links of family and tribe seems parochial and unfulfilling.

What the genome revolution has given us, though, is an even more important way to come to grips with who we are—a way to hold in our minds the extraordinary human diversity that exists today and has existed in our past. The problem of understanding the connections between self and the world is a central one for me, and has driven my lifelong interest in geography, history, and biology. Ironically for a person like myself, who is not at all religious, it is an example from the Bible that provides me with insight into how the genome revolution might be able to help solve this existential problem.

Every year on the holiday of Passover, Jews sit around the dinner table and recount the story of the Exodus from Egypt. The Passover holiday is important to Jews because it reminds them of their place in the world and encourages them to draw lessons about how they should behave. This narrative has been extraordinarily successful, as measured by the fact that it has sustained Jews in their identity for thousands of years as a minority living in foreign lands.

The Passover story begins with the myth of the patriarchs in ancient Israel: the first generation of Abraham and Sarah; the second of Isaac and Rebecca; the third of Jacob, Leah, Rachel, Bilhah, and Zilpah; and the fourth generation of twelve male children (the forefathers of the tribes of Israel) and a daughter, Dinah. These people are too removed from the huge populations of today to seem meaningfully connected to the present. The literary device that connects this ancient family to the multitudes that follow is Joseph, one of the sons of Jacob, who is sold by his brothers into slavery in Egypt, and who rises to a position of great power. When a famine strikes the land, the rest of the family also migrates to Egypt, where they are welcomed by Joseph despite the earlier crime they had committed against him. Four hundred years pass, and their descendants exponentially multiply into a nation numbering more than six hundred thousand military-age men and an even larger number of women and children. Under the leadership of Moses, they break their bonds of oppression, wander for dozens of years, and work out their code of laws. They then return to the Promised Land of their ancestors.

After reading the Passover story, Jews intuitively understand how within their population, numbering millions of people, they are related to each other and the past. The story allows Jews to think of those millions of coreligionists as direct relations—and to treat them with equal respect and seriousness even if they do not understand their exact relationships—to break out from the trap of thinking of the world from the perspective of the relatively small families we were raised in.

For me, the multitude of interconnected populations that have contributed to each of our genomes provide a similar narrative that helps me to understand my own place in the world and to avoid being daunted by the vast number of people in our species—the immensity of the human population numbering in the billions. The centrality of mixture in the history of our species, as revealed in just the last few years by the genome revolution, means that we are all interconnected and that we will all keep connecting with one another in the future. This narrative of connection allows me to feel Jewish even if I may not be descended from the matriarchs and patriarchs of the Bible. I feel American, even if I am not descended from indigenous Americans or the first European or African settlers. I speak English, a language not spoken by my ancestors a hundred years ago. I come from an intellectual tradition, the European Enlightenment, which is not that of my direct ancestors. I claim these as my own, even if they were not invented by my ancestors, even if I have no close genetic relationship to them. Our particular ancestors are not the point. The genome revolution provides us with a shared history that, if we pay proper attention, should give us an alternative to the evils of racism and nationalism, and make us realize that we are all entitled equally to our human heritage.

12

The Future of Ancient DNA

The Second Scientific Revolution in Archaeology

The first scientific revolution in archaeology began in 1949, when the chemist Willard Libby made a discovery that would transform the field forever and win him the Nobel Prize eleven years later.[1] He showed that by measuring the fraction of carbon atoms in ancient organic remains that carry fourteen nucleons instead of the more common twelve or thirteen, he could determine the date when the carbon first entered the food chain. On earth, the radioactive isotope carbon-14 is mostly formed through the bombardment of the atmosphere by cosmic rays, maintaining the proportion of all carbon atoms of this type at a level of about one part per trillion. During photosynthesis, plants pull carbon out of the atmosphere and change it into sugar. From there, it gets integrated into all the other molecules of life. After a living thing dies, half the carbon-14 atoms decay into nitrogen-14 within 5,730 years. This means the fraction of all carbon atoms in ancient remains that have fourteen nucleons decreases in a known way, enabling scientists to determine a date for when the carbon entered a living thing as long as the date is less than about fifty thousand years ago (beyond that, the fraction of carbon-14 is too low to make a measurement).

Radiocarbon dating transformed archaeology, making it possible to determine the true age of materials, going beyond what was possible by studying the layering of remains. The discoveries that archaeologists made were profound. In *Before Civilization: The Radiocarbon Revolution and Prehistoric Europe,* Colin Renfrew described how radiocarbon dating showed that human prehistory extended much further back in time than had previously been thought, and described how the radiocarbon revolution overturned the assumption that all major innovations in European prehistory were imports from the Near East.[2] While farming and writing were indeed of Near Eastern origin, innovations in metalworking and monumental constructions such as the building of megaliths like those at Stonehenge were not derived from ancient Egypt or Greece. These findings and many other discoveries about the true age of ancient remains sparked a new appreciation for indigenous cultures everywhere.

The penetration of radiocarbon dating into every aspect of archaeology is evident from the more than one hundred radiocarbon laboratories that provide dating to archaeologists as a service, and also from the fact that one of the basic skills serious archaeologists learn in graduate school is how to critically interpret radiocarbon dates. Radiocarbon dating has even changed archaeologists' yardstick for time. The ancient Chinese measured years since emperors ascended the throne; the Romans since the mythical foundation of their city; and the Jews since the date of the creation of the world according to the Bible. Almost everyone today denominates years before or after the supposed birth date of Jesus. For archaeologists, time is now measured as the number of radiocarbon decay years Before Present (BP), defined as 1950, the approximate year when Willard Libby discovered radiocarbon dating.

The radiocarbon revolution transformed the discipline of archaeology into one that by the 1960s was no longer only a branch of the humanities, and instead now had equally strong roots in the sciences, with a high standard of evidence now required to support claims.[3] Many additional scientific techniques were adopted by archaeologists in the period that followed, including flotation to identify ancient plant remains, and study of ratios of atomic isotopes beyond those of carbon to determine the types of foods peoples and animals ate and whether they moved across the landscape in their lifetimes. The rich new suite of scientific tools that archaeologists now had at their disposal made it possible for them to analyze the sites they excavated in ways that had not been possible for earlier generations of archaeologists, and to arrive at insights that were more reliable.

It is tempting to view ancient DNA as just one more new scientific technology that became available to archaeologists after the radiocarbon revolution, but that would be underestimating it. Prior to ancient DNA, archaeologists had hints of population movements based on the changes in the shapes of ancient skeletons and the types of artifacts people made, but these data were hard to interpret. But by sequencing whole genomes from ancient people, it is now possible to understand in exquisite detail how everyone is related.

The measure of a revolutionary technology is the rate at which it reveals surprises, and in this sense, ancient DNA is more revolutionary than any previous scientific technology for studying the past, including radiocarbon dating. A more apt analogy is the seventeenth-century invention of the light microscope, which made it possible to visualize the world of microbes and cells that no one before had even imagined. When a new instrument opens up vistas onto a world that has not previously been explored, everything it shows is new, and everything is a surprise. This is what is happening now with ancient DNA. It is providing definitive answers to questions about whether changes in the archaeological record reflect movements of people or cultural communication. Again and again, it is revealing findings that almost no one expected.

An Ancient DNA Atlas of Humanity

So far, the ancient DNA revolution has been highly Eurocentric. Of 551 published samples with genome-wide ancient DNA data as of late 2017, almost 90 percent are from West Eurasia. The focus on West Eurasia is a reflection of the fact that it is in Europe that most of the technology for ancient DNA analysis was developed, and it is in Europe that archaeologists have been studying their own backyards and collecting remains for the longest period of time. But the ancient DNA revolution is spreading, and has already produced several startling discoveries about human history outside of West Eurasia, most notably about the peopling of the Americas[4] and of the remote Pacific islands.[5] As technical improvements[6] have now made it possible to get ancient DNA from warm and even tropical places, I have no doubt that within the next decade, ancient DNA from central Asia, South Asia, East Asia, and Africa will reveal equally great surprises. The product of this effort will be an ancient DNA atlas of humanity, sampled densely through time and space. This will be a resource that I think will rival the first maps of the globe made between the fifteenth and nineteenth centuries in terms of its contribution to human knowledge. The atlas will not answer every question about population history, but it will provide a framework, a baseline to which we will always return when studying new archaeological sites.

There is every reason to expect an avalanche of major discoveries from ancient DNA over the coming years as this atlas is built. One of the key frontiers that has hardly been touched by ancient DNA is the period between four thousand years ago and the present. The great majority of samples studied so far have been older, but of course we know from the written record as well as from archaeological evidence that more recent times—the period of the development of writing, complex stratified societies, and empires—have been extraordinarily eventful. The corpus of ancient DNA data even in West Eurasia is like a highway overpass still under construction and ending in mid-air, not quite connecting the populations of the past to those of the present. Using DNA to address what happened in this period will surely add to what we know from other disciplines.

To bridge the last four thousand years, to connect the past to the present, it is not sufficient to simply collect ancient DNA data from recent periods. The statistical methods that have worked so well for studying the earlier periods break down when examining data from more recent times. In particular, the methods based on Four Population Tests owe their power to measuring the proportions of ancestry from populations that are highly differentiated—the very different ancestries act like tracer dyes whose changing proportions can be tracked. However, in Europe, where we have made most progress in the ancient DNA revolution so far, we know that by four thousand years ago, many populations were already highly similar in their ancestry composition to those of today.[7] For example, in Britain, we know that beginning after forty-five hundred years ago with people who buried their dead in association with wide-mouthed Bell Beaker pots, ancient Britons harbored a blend of ancestries very similar to that of present-day Britons.[8] Yet it would be a mistake to conclude from this that the people of Britain today are descended without mixture from the "Beaker folk." In fact, Britain's population has been transformed by multiple subsequent waves of migration of continental people who were genetically similar to the people associated with Beaker burials. New, more sensitive methods are needed to determine how much ancestry in Britain derives from later waves.

To address this challenge, statistical geneticists are developing a new class of methods that make it possible to track mixtures and migrations even of populations that are highly similar in their deep ancestral composition. The secret is to focus on the *recent* shared history of the analyzed populations instead of the *ancient* shared history. When a sufficiently large number of samples are analyzed together, it is possible to find segments of the genome in which pairs of individuals share close ancestors over the last approximately forty generations, and by focusing on these segments of the genome, we can learn what happened in human history over this time frame (roughly one thousand years).[9] With the small numbers of samples that have been available in ancient DNA studies so far, these methods have not been particularly useful because it is only the rare pair of individuals who are closely enough related to share identical long stretches of DNA. But as the number of individuals for whom we have ancient DNA increases, the number of pairs that we can analyze in order to detect relatedness increases according to the square of the number of samples. At the rate at which ancient DNA data are now being produced, it is reasonable to expect that within a few years, a single laboratory like mine will be producing genome-wide data from thousands of ancient people a year. This will make it possible to provide a detailed chronicle of how human populations have changed over recent millennia.

The power of this approach can already be seen in the 2015 study "The People of the British Isles," which sampled more than two thousand present-day individuals from the United Kingdom whose four grandparents were all born within eighty kilometers of one another.[10] The study found that the British population was very homogeneous by conventional measures. For example, the classic measure of genetic differentiation between two British populations is about one hundred times smaller than the same measurement of population differentiation comparing Europeans to East Asians. Despite the homogeneity, however, the authors were able to cluster the British population into seventeen crisply defined groups by searching for groups in which all pairs of individuals have elevated rates of recently shared genetic ancestors. Plotting the positions onto a map, they observed extraordinary genetic structuring, which has persisted despite the fact that people have moved back and forth continually over the British countryside over the past millennium, a process that would have been expected to homogenize the population. The boundaries of the clusters mark out the border between the southwestern counties of Devon and Cornwall; the Orkney Islands off the north coast of Scotland; a largely undifferentiated cluster crossing the Irish Sea reflecting the migration of Scottish Protestants to Northern Ireland within the last few centuries; and within Northern Ireland, two distinctive and barely mixing clusters, which surely correspond to the Protestant and Catholic populations, divided by religion and hundreds of years of enmity under British rule. The success of this analysis, performed only on present-day people, gives hope for extending the approach to samples that are more ancient. In my laboratory, we already have generated genome-wide data on more than three hundred ancient Britons. Coanalyzing them with present-day Britons, including those from the "People of the British Isles" study, we expect to be able to connect the dots between the past and the present in this one small part of the world.

Ancient DNA studies with large numbers of samples also offer the promise of being able to estimate human population sizes at different times in the past, a topic about which we have almost no reliable information from the period earlier than the invention of writing, but which is important for understanding not just human history and evolution but also economics and ecology. In a population of many hundreds of millions (such as the Han Chinese), a pair of randomly chosen people is expected to have few if any shared segments of DNA within the last forty generations because they descend from almost entirely different ancestors over this period. By contrast, in a small population (like the indigenous people of Little Andaman Island, who have a census size of fewer than one hundred), all pairs of individuals are closely related and will show evidence of relatedness through many shared segments of DNA. Measuring how related people are has been used to show, correctly, that the size of the population of England in the last few centuries has averaged many millions.[11] In ongoing work, Pier Palamara and I have demonstrated that the same approach can be used to show that early farmers from Anatolia of around eight thousand years ago were part of much larger populations than the hunter-gatherers from southern Sweden who were their contemporaries, as expected based on the higher densities that can be supported by agriculture. I have no doubt that applying this approach to ancient DNA will provide rich insight into how populations changed in size over time.

Ancient DNA's Promise for Revealing Human Biology

Ancient DNA in principle has just as much insight to offer about how human biology has changed over time as it does about human migrations and mixtures. And yet while the power of ancient DNA to reveal population transformations has been a runaway success, so far the insights into human biology have been limited. A key reason is that to track human biological change over time, it is important to be able to study how mutation frequencies change. But this requires hundreds of samples, and to date, the sample sizes of ancient DNA have been relatively small, just a handful from each cultural context. What will happen once we have genome-wide data from a thousand European farmers living shortly after the transition to agriculture? Comparing the results of a scan for recent natural selection in these individuals to the same scan performed in present-day Europeans should make it possible to understand whether the pace and nature of human adaptation has changed between preagricultural times and the time since the transition to agriculture. It might even be possible to determine whether natural selection has slowed down in the last century due to medical advances that allow individuals with genetic conditions that would have prevented them from surviving and having families to live and procreate. Examples of such medical conditions include poor eyesight, which can now be fully corrected with spectacles, or infertility, which can now be corrected by medical interventions, or cognitive challenges, which can now be controlled by medication and psychotherapy. It is possible that this change in natural selection is leading to a buildup of mutations contributing to altering these traits in the population.[12]

The power of ancient DNA to track the rate at which the frequencies of biologically important mutations have changed is important not just because it offers the possibility of tracking the evolution of specific traits, but also because it provides a previously unavailable tool that we can use to understand the fundamental principles of how natural selection proceeds. A central question in human evolutionary biology is whether human evolution typically proceeds by large changes in mutation frequencies at relatively small numbers of positions in the genome, as in the case of pigmentation, or by small changes in frequencies at a very large number of mutations, as in the case of height.[13] Understanding the relative importance of each type of adaptation is important, but addressing this question is made more challenging when the only tool available is analysis of people all of whom lived in a single window of time. Ancient DNA overcomes this obstacle—the time trap of only being able to study the present.

Ancient DNA research also reveals pathogen evolution. When grinding up human remains, we sometimes encounter DNA from microorganisms that were in an individual's bloodstream when he or she died and so were the likely cause of death. This approach proved that the bacterium *Yersinia pestis* was the cause of the fourteenth-to-seventeenth-century CE Black Death,[14] the sixth-to-eighth-century CE Justinianic plague of the Roman Empire,[15] and an endemic plague that was responsible for at least about 7 percent of deaths in skeletons from burials across the Eurasian steppe after around five thousand years ago.[16] Ancient pathogen studies have also revealed the history and origins of ancient leprosy,[17] tuberculosis,[18] and, in plants, the Irish potato famine.[19] Ancient DNA studies are now regularly obtaining material from the microbes that inhabit us, including from dental plaque and feces, providing information about the food our ancestors ate.[20] We are only just beginning to mine this new seam of information.

Taming the Wild West of the Ancient DNA Revolution

The speed at which the ancient DNA revolution is moving is exhilarating. The technology is evolving so quickly that many papers being published right now use methods that will be obsolete within a few years. Ancient DNA specialists are multiplying—for example, my own laboratory has already graduated three people who have founded their own ancient DNA laboratories. A major trend is specialization. The pioneers of ancient DNA spent a large portion of their time traveling the world to remote locations, talking with archaeologists and local officials, and bringing back unique remains that they have then analyzed in their molecular biology laboratories. Travel to exotic places and a gold rush to obtain key bones are central to this way of doing science. Some in the second generation of ancient DNA research have adopted this model. But others, including myself, travel far less, and instead spend most of our time developing expertise in improved laboratory techniques or statistical analysis, obtaining the samples we study through increasingly equal partnerships with archaeologists and anthropologists.

Ancient DNA laboratories will also become more specialized. At present, we who are working on ancient DNA have the privilege of doing research on populations from all over the world and from a wide range of times. We are like Robert Hooke turning his microscope to describe an extraordinary array of tiny objects in his book *Micrographia*, or like explorers in the late eighteenth century, sailing to every corner of the globe. But we have at best a superficial knowledge of the historical and archaeological and linguistic background of any topic we work on, and as knowledge grows, a deeper understanding of each region and the specific questions associated with it will be needed to make progress. Over the next two decades, I expect that ancient DNA specialists will be hired into every serious department of anthropology and archaeology, even history and biology. The professionals hired into these roles will be specialized in studying particular areas—for example, Southeast Asia or northeastern China—and their research will not flit from China to America to Europe to Africa as mine does today.

Ancient DNA will also go the way of specialization and even professionalization when it comes to setting up service laboratories, analogous to the service laboratories that exist for radiocarbon dating. Ancient DNA service laboratories will screen samples, generate genome-wide data, and provide reports that are easily interpretable, much like those currently provided by commercial personal ancestry testing companies. The reports will determine species, sex, and family relationships, and reveal how newly studied individuals relate to individuals for whom there is previously reported data. The researchers submitting the samples will receive an electronic copy of the data to use in any way they wish. The whole process shouldn't cost more than twice what radiocarbon dating does.

Service laboratories will proliferate, but researchers analyzing the data to study population history will never be entirely replaced. Archaeologists interested in learning about ancient populations using DNA will always need to partner with experts in genomics if they wish to use the technology to address any question that has subtlety. Getting information about sex, species, family relatedness, and ancestry outliers from ancient DNA will eventually be routine. But deeper scientific questions that can be accessed with ancient DNA data—such as how populations mixed and migrated, and how natural selection occurred over time—are unlikely ever to be addressed adequately through standardized reports.

The future for ancient DNA laboratories that I find appealing is based on a model that has emerged among radiocarbon dating laboratories. For example, the Oxford Radiocarbon Accelerator Unit processes large numbers of samples for a fee, and uses this income stream to support a factory that churns out routine dates and produces data more cheaply, efficiently, and at higher quality than would be possible if its scientists limited themselves to their own questions. But its scientists then piggyback on the juggernaut of the radiocarbon dating factory they have built to do cutting-edge science, such as the study led by Thomas Higham that clarified the record on the demise of Neanderthals in Europe, showing that they disappeared everywhere within a few thousand years of contact with modern humans.[21] This is also the model that I learned when I was a postdoctoral scientist at the Massachusetts Institute of Technology at one of the half dozen sequencing centers that carried out the brute-force work for the Human Genome Project, funded by large data production contracts from the U.S. National Institutes of Health. The center's leader, my supervisor, Eric Lander, also took advantage of the fact that he could turn the power of his sequencing center to address scientific problems that intrigued him. This is my model too: to build a factory, and then to commandeer it to answer deep questions about the past.

Out of Respect for Ancient Bones

I first went to Jerusalem when I was seven years old, taken there by my mother along with my older brother and younger sister. We stayed that summer and the next in an apartment that my grandfather owned in a poor, ultra-Orthodox neighborhood populated by men dressed in long black kaftans and women in layered modest dresses and headscarves. The boys attended morning-to-night religious schools, but on Friday afternoons before the Sabbath they were dismissed early and often joined political demonstrations. During the protests, they sometimes set fire to dumpsters and pelted policemen with stones. I remember watching the boys running, cloths pressed to faces, eyes streaming from the tear gas lobbed at them by the police.

Some of these protests were in response to excavations in the City of David, a site that spills down the hillside of the Temple Mount south of the Old City of Jerusalem, and covers much of the area that became the capital of Judaea after about three thousand years ago.

The protesters were upset that the excavations would disturb ancient Jewish graves, an ever-present possibility when digging in Israel. For the protesters, the opening of graves, whether by accident or for scientific investigation, was desecration.

What would those protesters think of what my laboratory is doing now, grinding through the bones of hundreds of ancient people every month? Perhaps they would not care much about samples from outside Israel, but I think the issue is more general, and I have found myself reflecting more and more about opening up the graves and sampling the remains of any ancient human. It is likely that many of the people whose bones we sample would not have wanted their remains to be used in this way.

One argument that some ancient DNA specialists and archaeologists have made is that most of the skeletons we are studying are from cultures so remote in time that they have no traceable connection to peoples of the present. This is the standard encoded in law in the U.S. Native American Graves Protection and Repatriation Act, which states that remains should be returned to Native American tribes when there is evidence of a cultural or biological connection to present-day peoples. However, this standard is now breaking down, as exemplified by the approximately 8,500-year-old Kennewick Man skeleton and the approximately 10,600-year-old Spirit Cave skeleton that are being returned to tribes despite having no clear cultural or genetic connections to specific groups living today.[22] As we study skeletons that draw ever closer in time to the present, it is important to think about the implications of modern claims on ancient samples. Ancient remains are the remains of real people whose physical integrity we should perhaps only violate if we have good reasons.

In 2016, I decided to ask a rabbi, in this case my mother's brother, for counsel. He is Orthodox, which means that he follows the intricate rules specified in the Jewish Oral Tradition. I had a hope that he might be open to my question, as he has also been an advocate of adapting Orthodox Judaism as much as possible to the modern world while abiding by the constraints of its fixed rules, a movement of inclusivity that has been called "Open Orthodoxy"—most recently, he set up a religious seminary to train women as Orthodox rabbis, a role from which women in that community had previously been excluded. I told him that in my lab we were grinding through the bones of ancient peoples, many of whom might not have wanted their remains to be disturbed, and that I felt I had not thought enough about this. He was obviously troubled, and asked me for some time to think. Afterward he came back with the judgment a rabbi gives to provide guidance when there is no precedent set by earlier decisions or judgments made by other rabbis. He said all human graves are sacrosanct, but there are mitigating circumstances that make it permissible to open graves as long as there is potential to promote understanding, to break down barriers between people.

The study of human variation has not always been a force for good. In Nazi Germany, someone with my expertise at interpreting genetic data would have been tasked with categorizing people by ancestry had that been possible with the science of the 1930s. But in our time, the findings from ancient DNA leave little solace for racist or nationalistic misinterpretation. In this field, the pursuit of truth for its own sake has overwhelmingly had the effect of exploding stereotypes, undercutting prejudice, and highlighting the connections among peoples not previously known to be related. I am optimistic that the direction of my work and that of my colleagues is to promote understanding, and I welcome our opportunity to do our best by the people, ancient and modern, whom we have been given the privilege to study. I see it as our role to midwife ancient DNA into a field that is not only the domain of geneticists, but also of archaeologists and the public—to realize its extraordinary potential to reveal who we are.

Notes on the Illustrations

Map sources. All maps were made with data from Natural Earth (http://www.natural earthdata.com/).

Figure 1. Contours in panel (a) are based on Fig. 2A of L. L. Cavalli-Sforza, P. Menozzi, and A. Piazza, "Demic Expansions and Human Evolution," *Science* 259 (1993): 639–46. Contours in panel (b) are based on interpolation of the numbers shown in Fig. 3 of W. Haak et al., "Massive Migration from the Steppe Was a Source for Indo-European Languages in Europe," *Nature* 522 (2015): 207–11. The interpolation was performed using the POPSutilities.R software of F. Jay et al., "Forecasting Changes in Population Genetic Structure of Alpine Plants in Response to Global Warming, *Molecular Ecology* (2012): 2354–68 and the parameter settings recommended in O. François, "Running Structure-like Population Genetic Analyses with R," June 2016, http://membres-timc.imag .fr/Olivier.Francois/tutoRstructure.pdf.

Figure 2. The plot shows the 3,748 unique individuals in the author's internal laboratory database as of November 19, 2017, broken down by the year when they became available.

Figure 4. The number of genealogical ancestors expected to have contributed DNA to a person living today is based on simulation results shared with the author by Graham Coop. The simulations were performed as described in G. Coop, "How Many Genetic Ancestors Do I Have," *gcbias* blog, November 11, 2013, https://gcbias.org/2013/11/11/how-does-your-number-of-genetic-ancestors -grow-back-over-time/.

Figure 5. The number of mutations in a given segment that separate the genome a person receives from his or her father and the one he or she receives from his or her mother can be used to estimate how much time has elapsed since the common ancestor at that location in the genome. Panel (2), which is based on the analyses reported in S. Mallick et al., "The Simons Genome Diversity

Project: 300 Genomes from 142 Diverse Populations," *Nature* 538 (2016): 201–6, shows the estimated times since the most recent shared ancestor averaged across 250 non-African genome pairs (solid line), and 44 sub-Saharan African genome pairs, measured at equally spaced locations in the DNA. Panel (3) shows the maximum estimated time at each location in the genome over 299 genome pairs and is based on analyses from the same study.

Figure 6. The approximate range of the Neanderthals is adapted from Fig. 1 of J. Krause et al., "Neanderthals in Central Asia and Siberia," *Nature* 449 (2007): 902–4.

Figure 7. The counts of shared mutations are based on the French-San-Neanderthal comparison in Table S48 of the Supplementary Online Materials of R. E. Green et al., "A Draft Sequence of the Neandertal Genome," *Science* 328 (2010): 710–22.

Figure 8. The illustration is based on the data in Fig. 2 of Q. Fu et al., "An Early Modern Human from Romania with a Recent Neanderthal Ancestor," *Nature* 524 (2015): 216–19.

Figure 9. This illustration replots the data shown in Fig. 2. of Q. Fu et al., "The Genetic History of Ice Age Europe," *Nature* 534 (2016): 200–5.

Figure 10. The pie chart data come from columns AJ and AK of Supplementary Table 2 of S. Mallick et al., "The Simons Genome Diversity Project: 300 Genomes from 142 Diverse Populations," *Nature* 538 (2016): 201–6. Each population is represented by an average of the individuals in that population. The proportion of archaic ancestry is expressed as a fraction of the maximum seen in any population in the dataset. Numbers less than 0.03 are set to 0 and numbers greater than 0.97 are set to 1. A subset of 47 populations is plotted to highlight the geographic coverage while reducing visual clutter.

Figure 13. This illustration represents the migrations in Europe described in Q. Fu et al., "The Genetic History of Ice Age Europe," *Nature* 534 (2016): 200–5. The ice extent is redrawn based on an online figure in "Extent of Ice Sheets in Europe," Map. *Encyclopaedia Britannica Online*, https://www.britannica.com/place/Scandinavian-Ice-Sheet?oasmId=54573.

Figure 14. Panel (a) is redrawn based on Extended Data Fig. 4 of W. Haak et al., "Massive Migration from the Steppe Was a Source for Indo-European Languages in Europe," *Nature* 522 (2015): 207–11. Panel (b) and its inset are adapted with permission from Fig. 1 and Fig. 2 of D. W. Anthony and D. Ringe, "The Indo-European Homeland from Linguistic and Archaeological Perspectives," *Annual Review of Linguistics* 1 (2015): 199–219.

Figure 15. The scatterplots in all three panels are based on the principal component analysis shown in Fig. 1b of I. Lazaridis et al., "Genetic Origins of the Minoans and Mycenaeans," *Nature* 548 (2017): 214–8. The x- and y-axes are rotated to roughly align genetic and geographic positions.

Figure 16. The pie charts are based on 180 Bell Beaker individuals for which there is enough ancient DNA data to make relatively precise estimates of steppe-related ancestry. The individuals are grouped by country within present-day Europe. The data are from a revised version of I. Olalde et al., "The Beaker Phenomenon and the Genomic Transformation of Northwest Europe," *bioRxiv* (2017): doi.org/10.1101/135962.

Figure 17. In panel (a), the South Asian Language family contours are redrawn based on a plot in *A Historical Atlas of South Asia*, ed. Joseph E. Schwartzberg (Oxford: Oxford University Press, 1992). In panel (b), the scatterplot is based on the principal component analysis in Fig. 3 of D. Reich et al., "Reconstructing Indian Population History," *Nature* 461 (2009): 489–94. The x- and y-axes are rotated to roughly align genetic and geographic positions.

Figure 18. The geographic contours and estimated dates for the spread of wheat and barley agriculture are drawn based on a sketch kindly provided by Dorian Fuller. The contours for the western half of the map follow Fig. 2 of F. Silva and M. Vander Linden, "Amplitude of Travelling Front as Inferred From 14C Predicts Levels of Genetic Admixture Among European Early Farmers," *Scientific Reports* 7 (2017): 11985.

Figure 19. The North American ice sheet and shoreline positions are derived from the figures on pages 380–83 of A. S. Dyke, "An Outline of North American Deglaciation with Emphasis on Central and Northern Canada," *Quaternary Glaciations—Extent and Chronology, Part II: North America*, ed. Jürgen Ehlers and Philip L. Gibbard (Amsterdam: Elsevier, 2004), 373–422. The Eurasian ice sheet positions are derived from Fig. 4 of H. Patton et al., "Deglaciation of the Eurasian Ice Sheet Complex," *Quaternary Science Reviews* 169 (2017): 148–72. The South American ice and shoreline positions are derived form Fig. 5.1 of D. J. Meltzer, "The Origins, Antiquity and Dispersal of the First Americans," in *The Human Past*, 4th Edition, ed. Chris Scarre (London: Thames and Hudson, expected early 2018), 149–71. The ancient Siberian shoreline is interpolated.

Figure 20. This illustration combines information from Fig. 2 of D. Reich et al., "Reconstructing Native American Population History," *Nature* 488 (2012): 370–74 and Fig. 5 of P. Flegontov et al., "Paleo-Eskimo Genetic Legacy Across North America," *bioRxiv* (2017): doi.org/10.1101.203018.

Figure 21. This illustration replots the data from Fig. 1 of P. Skoglund et al., "Genetic Evidence for Two Founding Populations of the Americas," *Nature* 525 (2015): 104–8.

Figure 23. The possible migration routes for early speakers of Tai-Kadai, Austroasiatic, and Austronesian languages are drawn based on Fig. 2 of J. Diamond and P. Bellwood, "Farmers and Their Languages: The First Expansions," *Science* 300 (2003): 597–603.

Figure 24. The ancient shoreline in panel (1) approximates the map in A. Cooper and C. Stringer, "Did the Denisovans Cross Wallace's Line?" *Science* 342 (2013): 321–23.

Figure 25. This illustration is based on Fig. 3D of P. Skoglund et al., "Reconstructing Prehistoric African Population Structure," *Cell* 171 (2017): 59–71.

Figure 26. The African language family contours approximate those shown in Fig. 3 of M. C. Campbell, J. B. Hirbo, J. P. Townsend, and S. A. Tishkoff, "The Peopling of the African Continent and the Diaspora into the New World," *Current Opinion in Genetics and Development* 29 (2014): 120–32. Possible migratory routes associated with the Bantu expansion are similar to those in Campbell et al., "The Peopling of the African Continent," but they also incorporate advice from Scott MacEachern and findings from subsequent genetic studies that suggest an expansion north of the tropical rainforest may not have contributed much of the ancestry of present-day Bantu speakers in East Africa (G. B. Busby et al., "Admixture into and Within Sub-Saharan Africa," *eLife* 5 (2016): e15266, and E. Patin et al., "Dispersals and Genetic Adaptation of Bantu-Speaking Populations in Africa and North America," *Science* 356 (2017): 543–46).

Figure 27. This illustration combines numbers from Fig. 2B and Fig. 2C of P. Skoglund et al., "Reconstructing Prehistoric African Population Structure," *Cell* 171 (2017): 59–71.

Figure 28. Adapted with permission from Fig. 2 of M. Karmin et al., "A Recent Bottleneck of Y Chromosome Diversity Coincides with a Global Change in Culture," *Genome Research* 25 (2015): 459–66.

Notes

Introduction

1. Luigi Luca Cavalli-Sforza, Paolo Menozzi, and Alberto Piazza, *The History and Geography of Human Genes* (Princeton, NJ: Princeton University Press, 1994).
2. Luigi Luca Cavalli-Sforza and Francesco Cavalli-Sforza, *The Great Human Diasporas: The History of Diversity and Evolution* (Reading, MA: Addison-Wesley, 1995).
3. N. A. Rosenberg et al., "Genetic Structure of Human Populations," *Science* 298 (2002): 2381–85.
4. P. Menozzi, A. Piazza, and L. L. Cavalli-Sforza, "Synthetic Maps of Human Gene Frequencies in Europeans," *Science* 201 (1978): 786–92; L. L. Cavalli-Sforza, P. Menozzi, and A. Piazza, "Demic Expansions and Human Evolution," *Science* 259 (1993): 639–46.
5. Albert J. Ammerman and Luigi Luca Cavalli-Sforza, *The Neolithic Transition and the Genetics of Populations in Europe* (Princeton, NJ: Princeton University Press, 1984).
6. J. Novembre and M. Stephens, "Interpreting Principal Component Analyses of Spatial Population Genetic Variation," *Nature Genetics* 40 (2008): 646–49.
7. O. François et al., "Principal Component Analysis Under Population Genetic Models of Range Expansion and Admixture," *Molecular Biology and Evolution* 27 (2010): 1257–68.
8. A. Keller et al., "New Insights into the Tyrolean Iceman's Origin and Phenotype as Inferred by Whole-Genome Sequencing," *Nature Communications* 3 (2012): 698; P. Skoglund et al., "Origins and Genetic Legacy of Neolithic Farmers and Hunter-Gatherers in Europe," *Science* 336 (2012): 466–69; I. Lazaridis et al., "Ancient Human Genomes Suggest Three Ancestral Populations for Present-Day Europeans," *Nature* 513 (2014): 409–13.
9. J. K. Pickrell and D. Reich, "Toward a New History and Geography of Human Genes Informed by Ancient DNA," *Trends in Genetics* 30 (2014): 377–89.
10. R. E. Green et al., "A Draft Sequence of the Neandertal Genome," *Science* 328 (2010): 710–22.

11. D. Reich et al., "Genetic History of an Archaic Hominin Group from Denisova Cave in Siberia," *Nature* 468 (2010): 1053–60.

12. M. Rasmussen et al., "Ancient Human Genome Sequence of an Extinct Palaeo-Eskimo," *Nature* 463 (2010): 757–62.

13. W. Haak et al., "Massive Migration from the Steppe Was a Source for Indo-European Languages in Europe," *Nature* 522 (2015): 207–11.

14. M. E. Allentoft et al., "Population Genomics of Bronze Age Eurasia," *Nature* 522 (2015): 167–72.

15. I. Mathieson et al., "Genome-Wide Patterns of Selection in 230 Ancient Eurasians," *Nature* 528 (2015): 499–503.

16. Q. Fu et al., "DNA Analysis of an Early Modern Human from Tianyuan Cave, China," *Proceedings of the National Academy of Sciences of the U.S.A.* 110 (2013): 2223–27.

17. H. Shang et al., "An Early Modern Human from Tianyuan Cave, Zhoukoudian, China," *Proceedings of the National Academy of Sciences of the U.S.A.* 104 (2007): 6573–78.

18. Haak et al., "Massive Migration."

19. I. Lazaridis et al., "Genomic Insights into the Origin of Farming in the Ancient Near East," *Nature* 536 (2016): 419–24.

20. P. Skoglund et al., "Genomic Insights into the Peopling of the Southwest Pacific," *Nature* 538 (2016): 510–13.

21. Lazaridis et al., "Ancient Human Genomes."

22. Pickrell and Reich, "Toward a New History."

1 How the Genome Explains Who We Are

1. J. D. Watson and F. H. Crick, "Molecular Structure of Nucleic Acids; a Structure for Deoxyribose Nucleic Acid," *Nature* 171 (1953): 737–38.

2. R. L. Cann, M. Stoneking, and A. C. Wilson, "Mitochondrial DNA and Human Evolution," *Nature* 325 (1987): 31–36.

3. Cann et al. "Mitochondrial DNA and Human Evolution."

4. Q. Fu et al., "A Revised Timescale for Human Evolution Based on Ancient Mitochondrial Genomes," *Current Biology* 23 (2013): 553–59.

5. D. E. Lieberman, B. M. McBratney, and G. Krovitz, "The Evolution and Development of Cranial Form in *Homo sapiens*," *Proceedings of the National Academy of Sciences of the U.S.A.* 99 (2002):1134–39. Richter et al., "The Age of the Hominin Fossils from Jebel Irhoud, Morocco, and the Origins of the Middle Stone Age," *Nature* 546 (2017): 293–96.

6. H. S. Groucutt et al., "Rethinking the Dispersal of *Homo sapiens* Out of Africa," *Evolutionary Anthropology* 24 (2015): 149–64.

7. C.-J. Kind et al., "The Smile of the Lion Man: Recent Excavations in Stadel Cave (Baden-Württemberg, South-Western Germany) and the Restoration of the Famous Upper Palaeolithic Figurine," *Quartär* 61 (2014): 129–45.

8. T. Higham et al., "The Timing and Spatiotemporal Patterning of Neanderthal Disappearance," *Nature* 512 (2014): 306–9.

9. Richard G. Klein and Blake Edgar, *The Dawn of Human Culture* (New York: Wiley, 2002).

10. J. Doebley, "Mapping the Genes That Made Maize," *Trends in Genetics* 8 (1992): 302–7.

11. S. McBrearty and A. S. Brooks, "The Revolution That Wasn't: A New Interpretation of the Origin of Modern Human Behavior," *Journal of Human Evolution* 39 (2000): 453–563.

12. C. S. L. Lai et al., "A Forkhead-Domain Gene Is Mutated in a Severe Speech and Language Disorder," *Nature* 413 (2001): 519–23.

13. W. Enard et al., "Molecular Evolution of *FOXP2*, a Gene Involved in Speech and Language," *Nature* 418 (2002): 869–72.

14. W. Enard et al., "A Humanized Version of *FOXP2* Affects Cortico-Basal Ganglia Circuits in Mice," *Cell* 137 (2009): 961–71.

15. J. Krause et al., "The Derived *FOXP2* Variant of Modern Humans Was Shared with Neandertals," *Current Biology* 17 (2007): 1908–12.

16. T. Maricic et al., "A Recent Evolutionary Change Affects a Regulatory Element in the Human *FOXP2* Gene," *Molecular Biology and Evolution* 30 (2013): 844–52.

17. S. Pääbo, "The Human Condition—a Molecular Approach," *Cell* 157 (2014): 216–26.

18. R. E. Green et al., "A Draft Sequence of the Neandertal Genome," *Science* 328 (2010): 710–22; K. Prüfer et al., "The Complete Genome Sequence of a Neanderthal from the Altai Mountains," *Nature* (2013): doi: 10.1038/nature 1288.

19. R. Lewin, "The Unmasking of Mitochondrial Eve," *Science* 238 (1987): 24–26.

20. A. Kong et al., "A High-Resolution Recombination Map of the Human Genome," *Nature Genetics* 31 (2002): 241–47.

21. "Descent of Elizabeth II from William I," Familypedia, http://familypedia .wikia.com/wiki/Descent_of_Elizabeth_II_from_William_I#Shorter_line_of_ descent.

22. S. Mallick et al., "The Simons Genome Diversity Project: 300 Genomes from 142 Diverse Populations," *Nature* 538 (2016): 201–6.

23. Green et al., "Draft Sequence."

24. H. Li and R. Durbin, "Inference of Human Population History from Individual Whole-Genome Sequences," *Nature* 475 (2011): 493–96.

25. Ibid.

26. S. Schiffels and R. Durbin, "Inferring Human Population Size and Separation History from Multiple Genome Sequences," *Nature Genetics* 46 (2014): 919–25.

27. Mallick et al., "Simons Genome Diversity Project."

28. I. Gronau et al., "Bayesian Inference of Ancient Human Demography from Individual Genome Sequences," *Nature Genetics* 43 (2011): 1031–34.

29. Mallick et al., "Simons Genome Diversity Project."

30. P. C. Sabeti et al., "Detecting Recent Positive Selection in the Human Genome from Haplotype Structure," *Nature* 419 (2002): 832–37; B. F. Voight, S. Kudaravalli, X. Wen, and J. K. Pritchard, "A Map of Recent Positive Selection in the Human Genome," *PLoS Biology* 4 (2006): e72.

31. K. M. Teshima, G. Coop, and M. Przeworski, "How Reliable Are Empirical Genomic Scans for Selective Sweeps?," *Genome Research* 16 (2006): 702–12.

32. R. D. Hernandez et al., "Classic Selective Sweeps Were Rare in Recent Human Evolution," *Science* 331 (2011): 920–24.

33. S. A. Tishkoff et al., "Convergent Adaptation of Human Lactase Persistence in Africa and Europe," *Nature Genetics* 38 (2006): 31–40.

34. M. C. Turchin et al., "Evidence of Widespread Selection on Standing Variation in Europe at Height-Associated SNPs," *Nature Genetics* 44 (2012): 1015–19.

35. I. Mathieson et al., "Genome-Wide Patterns of Selection in 230 Ancient Eurasians," *Nature* 528 (2015): 499–503.

36. Y. Field et al., "Detection of Human Adaptation During the Past 2000 Years," *Science* 354 (2016): 760–64.

37. D. Welter et al., "The NHGRI GWAS Catalog, a Curated Resource of SNP-Trait Associations," *Nucleic Acids Research* 42 (2014): D1001–6.

38. D. B. Goldstein, "Common Genetic Variation and Human Traits," *New England Journal of Medicine* 360 (2009): 1696–98.

39. A. Okbay et al., "Genome-Wide Association Study Identifies 74 Loci Associated with Educational Attainment," *Nature* 533 (2016): 539–42; M. T. Lo et al., "Genome-Wide Analyses for Personality Traits Identify Six Genomic Loci and Show Correlations with Psychiatric Disorders," *Nature Genetics* 49 (2017): 152–56; G. Davies et al., "Genome-Wide Association Study of Cognitive Functions and Educational Attainment in UK Biobank (N=112 151)," *Molecular Psychiatry* 21 (2016): 758–67.

2 Encounters with Neanderthals

1. Charles Darwin, *The Descent of Man, and Selection in Relation to Sex* (London: John Murray, 1871).

2. Erik Trinkaus, *The Shanidar Neanderthals* (New York: Academic Press, 1983).

3. D. Radovčić, A. O. Sršen, J. Radovčić, and D. W. Frayer, "Evidence for Neandertal Jewelry: Modified White-Tailed Eagle Claws at Krapina," *PLoS One* 10 (2015): e0119802.

4. J. Jaubert et al., "Early Neanderthal Constructions Deep in Bruniquel Cave in Southwestern France," *Nature* 534 (2016): 111–14.

5. W. L. Straus and A. J. E. Cave, "Pathology and the Posture of Neanderthal Man," *Quarterly Review of Biology* 32 (1957): 348–63.

6. William Golding, *The Inheritors* (London: Faber and Faber, 1955).

7. Jean M. Auel, *The Clan of the Cave Bear* (New York: Crown, 1980).

8. T. Higham et al., "The Timing and Spatiotemporal Patterning of Neanderthal Disappearance," *Nature* 512 (2014): 306–9.

9. T. Higham et al., "Chronology of the Grotte du Renne (France) and Implications for the Context of Ornaments and Human Remains Within the Châtelperronian," *Proceedings of the National Academy of Sciences of the U.S.A.* 107 (2010): 20234–39; O. Bar-Yosef and J.-G. Bordes, "Who Were the Makers of the Châtelperronian Culture?," *Journal of Human Evolution* 59 (2010): 586–93.

10. R. Grün et al., "U-series and ESR Analyses of Bones and Teeth Relating to the Human Burials from Skhul," *Journal of Human Evolution* 49 (2005): 316–34.

11. H. Valladas et al., "Thermo-Luminescence Dates for the Neanderthal Burial Site at Kebara in Israel," *Nature* 330 (1987): 159–60.

12. E. Trinkaus et al., "An Early Modern Human from the Peştera cu Oase, Romania," *Proceedings of the National Academy of Sciences of the U.S.A.* 100 (2003): 11231–36.

13. M. Krings et al., "Neandertal DNA Sequences and the Origin of Modern Humans," *Cell* 90 (1997): 19–30.

14. C. Posth et al., "Deeply Divergent Archaic Mitochondrial Genome Provides Lower Time Boundary for African Gene Flow into Neanderthals," *Nature Communications* 8 (2017): 16046.

15. Krings et al., "Neandertal DNA Sequences."

16. M. Currat and L. Excoffier, "Modern Humans Did Not Admix with Neanderthals During Their Range Expansion into Europe," *PLoS Biology* 2 (2004): e421; D. Serre et al., "No Evidence of Neandertal mtDNA Contribution to Early Modern Humans," *PLoS Biology* 2 (2004): e57; M. Nordborg, "On the Probability of Neanderthal Ancestry," *American Journal of Human Genetics* 63 (1998): 1237–40.

17. R. E. Green et al., "Analysis of One Million Base Pairs of Neanderthal DNA," *Nature* 444 (2006): 330–36.

18. J. D. Wall and S. K. Kim, "Inconsistencies in Neanderthal Genomic DNA Sequences," *PLoS Genetics* 3 (2007): 1862–66.

19. Krings et al., "Neandertal DNA Sequences."

20. S. Sankararaman et al., "The Date of Interbreeding Between Neandertals and Modern Humans," *PLoS Genetics* 8 (2012): e1002947.

21. P. Moorjani et al., "A Genetic Method for Dating Ancient Genomes Provides a Direct Estimate of Human Generation Interval in the Last 45,000 Years," *Proceedings of the National Academy of Sciences of the U.S.A.* 113 (2016): 5652–7.

22. G. Coop, "Thoughts On: The Date of Interbreeding Between Neandertals and Modern Humans," *Haldane's Sieve*, September 18, 2012, https://haldanessieve .org/2012/09/18/thoughts-on-neandertal-article/.

23. K. Prüfer et al., "The Complete Genome Sequence of a Neanderthal from the Altai Mountains," *Nature* (2013): doi: 10.1038/nature 12886.

24. Ibid.

25. Ibid; M. Meyer et al., "A High-Coverage Genome Sequence from an Archaic Denisovan Individual," *Science* 338 (2012): 222–26; J. D. Wall et al., "Higher Levels of Neanderthal Ancestry in East Asians Than in Europeans," *Genetics* 194 (2013): 199–209.

26. Q. Fu et al., "The Genetic History of Ice Age Europe," *Nature* 534 (2016): 200–5.

27. I. Lazaridis et al., "Genomic Insights into the Origin of Farming in the Ancient Near East," *Nature* 536 (2016): 419–24.

28. Trinkaus et al., "An Early Modern Human."

29. Q. Fu et al., "An Early Modern Human from Romania with a Recent Neanderthal Ancestor," *Nature* 524 (2015): 216–19.

30. N. Teyssandier, F. Bon, and J.-G. Bordes, "Within Projectile Range: Some Thoughts on the Appearance of the Aurignacian in Europe," *Journal of Anthropological Research* 66 (2010): 209–29; P. Mellars, "Archeology and the Dispersal of Modern Humans in Europe: Deconstructing the 'Aurignacian,'" *Evolutionary Anthropology* 15 (2006): 167–82.

31. M. Currat and L. Excoffier, "Strong Reproductive Isolation Between Humans and Neanderthals Inferred from Observed Patterns of Introgression," *Proceedings of the National Academy of Sciences of the U.S.A.* 108 (2011): 15129–34.

32. S. Sankararaman et al., "The Genomic Landscape of Neanderthal Ancestry in Present-Day Humans," *Nature* 507 (2014): 354–57; B. Vernot and J. M.

Akey, "Resurrecting Surviving Neandertal Lineages from Modern Human Genomes," *Science* 343 (2014): 1017-21.

33. N. Patterson et al., "Genetic Evidence for Complex Speciation of Humans and Chimpanzees," *Nature* 441 (2006): 1103–8.

34. Ibid; R. Burgess and Z. Yang, "Estimation of Hominoid Ancestral Population Sizes Under Bayesian Coalescent Models Incorporating Mutation Rate Variation and Sequencing Errors," *Molecular Biology and Evolution* 25 (2008): 1975–94.

35. J. A. Coyne and H. A. Orr, "Two Rules of Speciation," in *Speciation and Its Consequences*, ed. Daniel Otte and John A. Endler (Sunderland, MA: Sinauer Associates, 1989), 180–207.

36. P. K. Tucker et al., "Abrupt Cline for Sex-Chromosomes in a Hybrid Zone Between Two Species of Mice," *Evolution* 46 (1992): 1146–63.

37. H. Li and R. Durbin, "Inference of Human Population History from Individual Whole-Genome Sequences," *Nature* 475 (2011): 493–96.

38. T. Mailund et al., "A New Isolation with Migration Model Along Complete Genomes Infers Very Different Divergence Processes Among Closely Related Great Ape Species," *PLoS Genetics* 8 (2012): e1003125.

39. J. Y. Dutheil et al., "Strong Selective Sweeps on the X Chromosome in the Human-Chimpanzee Ancestor Explain Its Low Divergence," *PLoS Genetics* 11 (2015): e1005451.

40. Sankararaman et al., "Genomic Landscape"; B. Jégou et al., "Meiotic Genes Are Enriched in Regions of Reduced Archaic Ancestry," *Molecular Biology and Evolution* 34 (2017): 1974–80.

41. Q. Fu et al., "Ice Age Europe."

42. I. Juric, S. Aeschbacher, and G. Coop, "The Strength of Selection Against Neanderthal Introgression," *PLoS Genetics* 12 (2016): e1006340; K. Harris and R. Nielsen, "The Genetic Cost of Neanderthal Introgression," *Genetics* 203 (2016): 881–91.

43. G. Bhatia et al., "Genome-Wide Scan of 29,141 African Americans Finds No Evidence of Directional Selection Since Admixture," *American Journal of Human Genetics* 95 (2014): 437–44.

44. Johann G. Fichte, *Grundlage der gesamten Wissenschaftslehre* (Jena, Germany: Gabler, 1794).

3 Ancient DNA Opens the Floodgates

1. J. Krause et al., "Neanderthals in Central Asia and Siberia," *Nature* 449 (2007): 902–4.

2. J. Krause et al., "The Complete Mitochondrial DNA Genome of an Unknown Hominin from Southern Siberia," *Nature* 464 (2010): 894-97.

3. C. Posth et al., "Deeply Divergent Archaic Mitochondrial Genome Provides Lower Time Boundary for African Gene Flow into Neanderthals," *Nature Communications* 8 (2017): 16046.

4. Krause et al., "Unknown Hominin."

5. D. Reich et al., "Genetic History of an Archaic Hominin Group from Denisova Cave in Siberia," *Nature* 468 (2010): 1053–60.

6. K. Prüfer et al., "The Complete Genome Sequence of a Neanderthal from the Altai Mountains," *Nature* (2013): doi: 10.1038/nature 12886.

7. Jerry A. Coyne and H. Allen Orr, *Speciation* (Sunderland, MA: Sinauer Associates, 2004).

8. S. Sankararaman, S. Mallick, N. Patterson, and D. Reich, "The Combined Landscape of Denisovan and Neanderthal Ancestry in Present-Day Humans," *Current Biology* 26 (2016): 1241–47.

9. P. Moorjani et al., "A Genetic Method for Dating Ancient Genomes Provides a Direct Estimate of Human Generation Interval in the Last 45,000 Years," *Proceedings of the National Academy of Sciences of the U.S.A.* 113 (2016): 5652–7.

10. Sankararaman et al., "Combined Landscape."

11. D. Reich et al., "Denisova Admixture and the First Modern Human Dispersals into Southeast Asia and Oceania," *American Journal of Human Genetics* 89 (2011): 516–28.

12. Q. Fu et al., "DNA Analysis of an Early Modern Human from Tianyuan Cave, China," *Proceedings of the National Academy of Sciences of the U.S.A.* 110 (2013): 2223–27; M. Yang et al., "40,000-Year-Old Individual from Asia Provides Insight into Early Population Structure in Eurasia," *Current Biology* 27 (2017): 3202–8.

13. Prüfer et al., "Complete Genome."

14. C. B. Stringer and I. Barnes, "Deciphering the Denisovans," *Proceedings of the National Academy of Sciences of the U.S.A.* 112 (2015): 15542–43.

15. G. A. Wagner et al., "Radiometric Dating of the Type-Site for *Homo Heidelbergensis* at Mauer, Germany," *Proceedings of the National Academy of Sciences of the U.S.A.* 107 (2010): 19726–30.

16. C. Stringer, "The Status of *Homo heidelbergensis* (Schoetensack 1908)," *Evolutionary Anthropology* 21 (2012): 101–7.

17. A. Brumm et al., "Age and Context of the Oldest Known Hominin Fossils from Flores," *Nature* 534 (2016): 249–53.

18. Reich et al., "Denisova Admixture."

19. Prüfer et al., "Complete Genome."

20. Ibid.; Sankararaman et al., "Combined Landscape."

21. E. Huerta-Sánchez et al., "Altitude Adaptation in Tibetans Caused by Introgression of Denisovan-like DNA," *Nature* 512 (2014): 194–97.

22. F. H. Chen et al., "Agriculture Facilitated Permanent Human Occupation of the Tibetan Plateau After 3600 B.P.," *Science* 347 (2015): 248–50.

23. S. Sankararaman et al., "The Genomic Landscape of Neanderthal Ancestry in Present-Day Humans," *Nature* 507 (2014): 354–57; B. Vernot and J. M. Akey, "Resurrecting Surviving Neandertal Lineages from Modern Human Genomes," *Science* 343 (2014): 1017–21.

24. Prüfer et al., "Complete Genome."

25. G. P. Rightmire, "*Homo erectus* and Middle Pleistocene Hominins: Brain Size, Skull Form, and Species Recognition," *Journal of Human Evolution* 65 (2013): 223–52.

26. M. Martinón-Torres et al., "Dental Evidence on the Hominin Dispersals During the Pleistocene," *Proceedings of the National Academy of Sciences of the U.S.A.* 104 (2007): 13279–82; M. Martinón-Torres, R. Dennell, and J. M. B. de Castro, "The Denisova Hominin Need Not Be an Out of Africa Story," *Journal*

of Human Evolution 60 (2011): 251–55; J. M. B. de Castro and M. Martinón-Torres, "A New Model for the Evolution of the Human Pleistocene Populations of Europe," *Quaternary International* 295 (2013): 102–12.

27. De Castro and Martinón-Torres, "A New Model."

28. J. L. Arsuaga et al., "Neandertal Roots: Cranial and Chronological Evidence from Sima de los Huesos," *Science* 344 (2014): 1358–63; M. Meyer et al., "A Mitochondrial Genome Sequence of a Hominin from Sima de los Huesos," *Nature* 505 (2014): 403–6.

29. M. Meyer et al., "Nuclear DNA Sequences from the Middle Pleistocene Sima de los Huesos Hominins," *Nature* 531 (2016): 504–7.

30. Meyer et al., "A Mitochondrial Genome"; Meyer et al., "Nuclear DNA Sequences."

31. Krause et al., "Unknown Hominin"; Reich et al., "Genetic History."

32. Posth et al., "Deeply Divergent Archaic."

33. Ibid.

34. Prüfer et al., "Complete Genome."

35. S. McBrearty and A. S. Brooks, "The Revolution That Wasn't: A New Interpretation of the Origin of Modern Human Behavior," *Journal of Human Evolution* 39 (2000): 453–563.

36. M. Kuhlwilm et al., "Ancient Gene Flow from Early Modern Humans into Eastern Neanderthals," *Nature* 530 (2016): 429–33.

4 Humanity's Ghosts

1. Charles R. Darwin, *On the Origin of Species by Means of Natural Selection, or the Preservation of Favoured Races in the Struggle for Life* (London: John Murray, 1859).

2. C. Becquet et al., "Genetic Structure of Chimpanzee Populations," *PLoS Genetics* 3 (2007): e66.

3. R. E. Green et al., "A Draft Sequence of the Neandertal Genome," *Science* 328 (2010): 710–22.

4. N. J. Patterson et al., "Ancient Admixture in Human History," *Genetics* 192 (2012): 1065–93.

5. Ernst Mayr, *Systematics and the Origin of Species from the Viewpoint of a Zoologist* (New York: Columbia University Press, 1942).

6. J. K. Pickrell and D. Reich, "Toward a New History and Geography of Human Genes Informed by Ancient DNA," *Trends in Genetics* 30 (2014): 377–89.

7. A. R. Templeton, "Biological Races in Humans," *Studies in History and Philosophy of Biological and Biomedical Science* 44 (2013): 262–71.

8. M. Raghavan et al., "Upper Palaeolithic Siberian Genome Reveals Dual Ancestry of Native Americans," *Nature* 505 (2014): 87–91.

9. I. Lazaridis et al., "Ancient Human Genomes Suggest Three Ancestral Populations for Present-Day Europeans," *Nature* 513 (2014): 409–13.

10. I. Lazaridis et al., "Genomic Insights into the Origin of Farming in the Ancient Near East," *Nature* 536 (2016): 419–24.

11. Ibid.

12. F. Broushaki et al., "Early Neolithic Genomes from the Eastern Fertile Cres-

cent," *Science* 353 (2016): 499–503; E. R. Jones et al., "Upper Palaeolithic Genomes Reveal Deep Roots of Modern Eurasians," *Nature Communications* 6 (2015): 8912.

13. B. M. Henn et al., "Genomic Ancestry of North Africans Supports Back-to-Africa Migrations," *PLoS Genetics* 8 (2012): e1002397.

14. Lazaridis et al., "Genomic Insights."

15. O. Bar-Yosef, "Pleistocene Connections Between Africa and Southwest Asia: An Archaeological Perspective," *African Archaeological Review* 5 (1987): 29–38.

16. Lazaridis et al., "Genomic Insights."

17. Lazaridis et al., "Ancient Human Genomes."

18. Q. Fu et al., "The Genetic History of Ice Age Europe," *Nature* 534 (2016): 200–5.

19. Q. Fu et al., "Genome Sequence of a 45,000-Year-Old Modern Human from Western Siberia," *Nature* 514 (2014): 445–49.

20. Q. Fu et al., "An Early Modern Human from Romania with a Recent Neanderthal Ancestor," *Nature* 524 (2015): 216–19.

21. F. G. Fedele, B. Giaccio, and I. Hajdas, "Timescales and Cultural Process at 40,000 BP in the Light of the Campanian Ignimbrite Eruption, Western Eurasia," *Journal of Human Evolution* 55 (2008): 834–57; A. Costa et al., "Quantifying Volcanic Ash Dispersal and Impact of the Campanian Ignimbrite Super-Eruption," *Geophysical Research Letters* 39 (2012): L10310.

22. Fedele et al., "Timescales and Cultural Process."

23. A. Seguin-Orlando et al., "Genomic Structure in Europeans Dating Back at Least 36,200 Years," *Science* 346 (2014): 1113–18.

24. Fu et al., "Ice Age Europe."

25. Andreas Maier, *The Central European Magdalenian: Regional Diversity and Internal Variability* (Dordrecht, The Netherlands: Springer, 2015).

26. Fu et al., "Ice Age Europe."

27. N. A. Rosenberg et al., "Clines, Clusters, and the Effect of Study Design on the Inference of Human Population Structure," *PLoS Genetics* 1 (2005): e70; G. Coop et al., "The Role of Geography in Human Adaptation," *PLoS Genetics* 5 (2009): e1000500.

28. Q. Fu et al., "DNA Analysis of an Early Modern Human from Tianyuan Cave, China," *Proceedings of the National Academy of Sciences of the U.S.A.* 110 (2013): 2223–27.

29. Fu et al., "Recent Neanderthal Ancestor"; W. Haak et al., "Massive Migration from the Steppe Was a Source for Indo-European Languages in Europe," *Nature* 522 (2015): 207–11.

30. R. Pinhasi et al., "Optimal Ancient DNA Yields from the Inner Ear Part of the Human Petrous Bone," *PLoS One* 10 (2015): e0129102.

31. Lazaridis et al., "Genomic Insights."

32. Ibid.; Broushaki et al., "Early Neolithic Genomes."

33. I. Olalde et al., "Derived Immune and Ancestral Pigmentation Alleles in a 7,000-Year-Old Mesolithic European," *Nature* 507 (2014): 225–28.

34. I. Mathieson et al., "Genome-Wide Patterns of Selection in 230 Ancient Eurasians," *Nature* 528 (2015): 499–503.

35. I. Mathieson et al., "The Genomic History of Southeastern Europe," *bioRxiv* (2017): doi.org/10.1101/135616.

36. Haak et al., "Massive Migration"; M. E. Allentoft et al., "Population Genomics of Bronze Age Eurasia," *Nature* 522 (2015): 167–72.
37. Templeton, "Biological Races."

5 The Making of Modern Europe

1. B. Bramanti et al., "Genetic Discontinuity Between Local Hunter-Gatherers and Central Europe's First Farmers," *Science* 326 (2009): 137–40.
2. A. Keller et al., "New Insights into the Tyrolean Iceman's Origin and Phenotype as Inferred by Whole-Genome Sequencing," *Nature Communications* 3 (2012): 698.
3. W. Muller et al., "Origin and Migration of the Alpine Iceman," *Science* 302 (2003): 862–66.
4. P. Skoglund et al., "Origins and Genetic Legacy of Neolithic Farmers and Hunter-Gatherers in Europe," *Science* 336 (2012): 466–69.
5. Albert J. Ammerman and Luigi Luca Cavalli-Sforza, *The Neolithic Transition and the Genetics of Populations in Europe* (Princeton, NJ: Princeton University Press, 1984).
6. N. J. Patterson et al., "Ancient Admixture in Human History," *Genetics* 192 (2012): 1065–93.
7. M. Raghavan et al., "Upper Palaeolithic Siberian Genome Reveals Dual Ancestry of Native Americans," *Nature* (2013): doi: 10.1038/nature12736.
8. I. Lazaridis et al., "Ancient Human Genomes Suggest Three Ancestral Populations for Present-Day Europeans," *Nature* 513 (2014): 409–13.
9. C. Gamba et al., "Genome Flux and Stasis in a Five Millennium Transect of European Prehistory," *Nature Communications* 5 (2014): 5257; M. E. Allentoft et al., "Population Genomics of Bronze Age Eurasia," *Nature* 522 (2015): 167–72; W. Haak et al., "Massive Migration from the Steppe Was a Source for Indo-European Languages in Europe," *Nature* 522 (2015): 207–11; I. Mathieson et al., "Genome-Wide Patterns of Selection in 230 Ancient Eurasians," *Nature* 528 (2015): 499–503.
10. Luigi Luca Cavalli-Sforza, Paolo Menozzi, and Alberto Piazza, *The History and Geography of Human Genes* (Princeton, NJ: Princeton University Press, 1994).
11. Haak et al., "Massive Migration"; Mathieson et al., "Genome-Wide Patterns."
12. Q. Fu et al., "The Genetic History of Ice Age Europe," *Nature* 534 (2016): 200–5.
13. I. Mathieson, "The Genomic History of Southeastern Europe," *bioRxiv* (2017): doi.org/10.1101/135616.
14. K. Douka et al., "Dating Knossos and the Arrival of the Earliest Neolithic in the Southern Aegean," *Antiquity* 91 (2017): 304–21.
15. Haak et al., "Massive Migration"; M. Lipson et al., "Parallel Palaeogenomic Transects Reveal Complex Genetic History of Early European Farmers," *Nature* 551 (2017): 368–72.
16. Colin Renfrew, *Before Civilization: The Radiocarbon Revolution and Prehistoric Europe* (London: Jonathan Cape, 1973).
17. Marija Gimbutas, *The Prehistory of Eastern Europe, Part I: Mesolithic, Neolithic and Copper Age Cultures in Russia and the Baltic Area* (American School of Prehistoric

Research, Harvard University, Bulletin No. 20) (Cambridge, MA: Peabody Museum, 1956).

18. David W. Anthony, *The Horse, the Wheel, and Language: How Bronze-Age Riders from the Eurasian Steppes Shaped the Modern World* (Princeton, NJ: Princeton University Press, 2007).

19. Ibid.

20. Ibid.

21. Haak et al., "Massive Migration."

22. Ibid.; I. Lazaridis et al., "Genomic Insights into the Origin of Farming in the Ancient Near East," *Nature* 536 (2016): 419–24.

23. M. Ivanova, "Kaukasus Und Orient: Die Entstehung des 'Maikop-Phänomens' im 4. Jahrtausend v. Chr.," *Praehistorische Zeitschrift* 87 (2012): 1–28.

24. Haak et al., "Massive Migration"; Allentoft et al., "Bronze Age Eurasia."

25. Ibid.

26. G. Kossinna, "Die Deutsche Ostmark: Ein Heimatboden der Germanen," *Berlin* (1919).

27. B. Arnold, "The Past as Propaganda: Totalitarian Archaeology in Nazi Germany," *Antiquity* 64 (1990): 464–78.

28. H. Härke, "The Debate on Migration and Identity in Europe," *Antiquity* 78 (2004): 453–56.

29. V. Heyd, "Kossinna's Smile," *Antiquity* 91 (2017): 348–59; M. Vander Linden, "Population History in Third-Millennium-BC Europe: Assessing the Contribution of Genetics," *World Archaeology* 48 (2016): 714–28; N. N. Johannsen, G. Larson, D. J. Meltzer, and M. Vander Linden, "A Composite Window into Human History," *Science* 356 (2017): 1118–20.

30. Vere Gordon Childe, *The Aryans: A Study of Indo-European Origins* (London and New York: K. Paul, Trench, Trubner and Co. and Alfred A. Knopf, 1926).

31. Härke, "Debate on Migration and Identity."

32. Peter Bellwood, *First Migrants: Ancient Migration in Global Perspective* (Chichester, West Sussex, UK / Malden, MA: Wiley-Blackwell, 2013).

33. Colin McEvedy and Richard Jones, *Atlas of World Population History* (Harmondsworth, Middlesex, UK: Penguin, 1978).

34. K. Kristiansen, "The Bronze Age Expansion of Indo-European Languages: An Archaeological Model," in *Becoming European: The Transformation of Third Millennium Northern and Western Europe*, ed. Christopher Prescott and Håkon Glørstad (Oxford: Oxbow Books, 2011), 165–81.

35. S. Rasmussen et al., "Early Divergent Strains of *Yersinia pestis* in Eurasia 5,000 Years Ago," *Cell* 163 (2015): 571–82.

36. A. P. Fitzpatrick, *The Amesbury Archer and the Boscombe Bowmen: Bell Beaker Burials at Boscombe Down, Amesbury, Wiltshire* (Salisbury, UK: Wessex Archaeology Reports, 2011).

37. I. Olalde et al., "The Beaker Phenomenon and the Genomic Transformation of Northwest Europe," *bioRxiv* (2017): doi.org/10.1101/135962.

38. L. M. Cassidy et al., "Neolithic and Bronze Age Migration to Ireland and Establishment of the Insular Atlantic Genome," *Proceedings of the National Academy of Sciences of the U.S.A.* 113 (2016): 368–73.

39. Colin Renfrew, *Archaeology and Language: The Puzzle of Indo-European Origins* (Cambridge: Cambridge University Press, 1997).

40. Ibid.

41. P. Bellwood, "Human Migrations and the Histories of Major Language Families," in *The Global Prehistory of Human Migration* (Chichester, UK, and Malden, MA: Wiley-Blackwell, 2013), 87–95.

42. Renfrew, *Archaeology and Language*; Peter Bellwood, *First Farmers: The Origins of Agricultural Societies* (Malden, MA: Blackwell, 2005).

43. Haak et al., "Massive Migration"; Allentoft et al., "Bronze Age Eurasia."

44. D. W. Anthony and D. Ringe, "The Indo-European Homeland from Linguistic and Archaeological Perspectives," *Annual Review of Linguistics* 1 (2015): 199–219.

45. Léon Poliakov, *The Aryan Myth: A History of Racist and Nationalist Ideas in Europe* (New York: Basic Books, 1974).

6 The Collision That Formed India

1. *The Rigveda*, trans. Stephanie W. Jamison and Joel P. Brereton (Oxford: Oxford University Press, 2014), hymns 1.33, 1.53, 2.12, 3.30, 3.34, 4.16, and 4.28.

2. M. Witzel, "Early Indian History: Linguistic and Textual Parameters," in *The Indo-Aryans of Ancient South Asia: Language, Material Culture and Ethnicity*, ed. George Erdosy (Berlin: Walter de Gruyter, 1995), 85–125.

3. Rita P. Wright, *The Ancient Indus: Urbanism, Economy, and Society* (Cambridge: Cambridge University Press, 2010); Gregory L. Possehl, *The Indus Civilization: A Contemporary Perspective* (Lanham, MD: AltaMira Press, 2002).

4. Ibid.

5. Asko Parpola, *Deciphering the Indus Script* (Cambridge: Cambridge University Press, 1994); S. Farmer, R. Sproat, and M. Witzel, "The Collapse of the Indus-Script Thesis: The Myth of a Literate Harappan Civilization," *Electronic Journal of Vedic Studies* 11 (2004): 19–57.

6. Richard H. Meadow, ed., *Harappa Excavations 1986–1990: A Multidisciplinary Approach to Third Millennium Urbanism* (Madison, WI: Prehistory Press, 1991); A. Lawler, "Indus Collapse: The End or the Beginning of an Asian Culture?," *Science* 320 (2008): 1281–83.

7. Jaan Puhvel, *Comparative Mythology* (Baltimore: Johns Hopkins University Press, 1987).

8. Wright, *The Ancient Indus*; Possehl, *The Indus Civilization*.

9. Alfred Rosenberg, *The Myth of the Twentieth Century: An Evaluation of the Spiritual-Intellectual Confrontations of Our Age*, trans. Vivian Bird (Torrance, CA: Noontide Press, 1982).

10. Léon Poliakov, *The Aryan Myth: A History of Racist and Nationalist Ideas in Europe* (New York: Basic Books, 1974).

11. B. Arnold, "The Past as Propaganda: Totalitarian Archaeology in Nazi Germany," *Antiquity* 64 (1990): 464–78.

12. Bryan Ward-Perkis, *The Fall of Rome and the End of Civilization* (Oxford: Oxford University Press, 2005).

13. Peter Bellwood, *First Farmers: The Origins of Agricultural Societies* (Malden, MA: Blackwell, 2005).

14. Ibid.

15. M. Witzel, "Substrate Languages in Old Indo-Aryan (Rgvedic, Middle and Late Vedic)," *Electronic Journal of Vedic Studies* 5 (1999): 1–67.

16. K. Thangaraj et al., "Reconstructing the Origin of Andaman Islanders," *Science* 308 (2005): 996; K. Thangaraj et al., "*In situ* Origin of Deep Rooting Lineages of Mitochondrial Macrohaplogroup 'M' in India," *BMC Genomics* 7 (2006): 151.

17. R. S. Wells et al., "The Eurasian Heartland: A Continental Perspective on Y-chromosome Diversity," *Proceedings of the National Academy of Sciences of the U.S.A.* 98 (2001): 10244–49; M. Bamshad et al., "Genetic Evidence on the Origins of Indian Caste Populations," *Genome Research* 11 (2001): 994–1004; I. Thanseem et al., "Genetic Affinities Among the Lower Castes and Tribal Groups of India: Inference from Y Chromosome and Mitochondrial DNA," *BMC Genetics* 7 (2006): 42.

18. Thangaraj et al., "Andaman Islanders."

19. D. Reich et al., "Reconstructing Indian Population History," *Nature* 461 (2009): 489–94.

20. R. E. Green et al., "A Draft Sequence of the Neandertal Genome," *Science* 328 (2010): 710–22.

21. Thangaraj et al., "Deep Rooting Lineages."

22. Reich et al., "Reconstructing Indian Population History"; P. Moorjani et al., "Genetic Evidence for Recent Population Mixture in India," *American Journal of Human Genetics* 93 (2013): 422–38.

23. Ibid.

24. Irawati Karve, *Hindu Society—An Interpretation* (Pune, India: Deccan College Post Graduate and Research Institute, 1961).

25. P. A. Underhill et al., "The Phylogenetic and Geographic Structure of Y-Chromosome Haplogroup R1a," *European Journal of Human Genetics* 23 (2015): 124–31.

26. S. Perur, "The Origins of Indians: What Our Genes Are Telling Us," *Fountain Ink*, December 3, 2013, http://fountainink.in/?p=4669&all=1.

27. K. Bryc et al., "The Genetic Ancestry of African Americans, Latinos, and European Americans Across the United States," *American Journal of Human Genetics* 96 (2015): 37–53.

28. L. G. Carvajal-Carmona et al., "Strong Amerind/White Sex Bias and a Possible Sephardic Contribution Among the Founders of a Population in Northwest Colombia," *American Journal of Human Genetics* 67 (2000): 1287–95; G. Bedoya et al., "Admixture Dynamics in Hispanics: A Shift in the Nuclear Genetic Ancestry of a South American Population Isolate," *Proceedings of the National Academy of Sciences of the U.S.A.* 103 (2006): 7234–39.

29. Moorjani et al., "Recent Population Mixture."

30. Ibid.

31. Romila Thapar, *Early India: From the Origins to AD 1300* (Berkeley: University of California Press, 2002); Karve, *Hindu Society;* Susan Bayly, *Caste, Society and Politics in India from the Eighteenth Century to the Modern Age* (Cambridge: Cambridge University Press, 1999); M. N. Srinivas, *Caste in Modern India and Other Essays* (Bombay: Asia Publishing House, 1962); Louis Dumont, *Homo Hierarchicus: The Caste System and Its Implications* (Chicago: University of Chicago Press, 1980).

32. Kumar Suresh Singh, *People of India: An Introduction* (People of India National Series) (New Delhi: Oxford University Press, 2002); K. C. Malhotra and T. S. Vasulu, "Structure of Human Populations in India," in *Human Population Genet-*

ics: A Centennial Tribute to J. B. S. Haldane, ed. Partha P. Majumder (New York: Plenum Press, 1993), 207–34.

33. Karve, "Hindu Society."

34. Ibid.

35. Nicholas B. Dirks, *Castes of Mind: Colonialism and the Making of Modern India* (Princeton, NJ: Princeton University Press, 2001); N. Boivin, "Anthropological, Historical, Archaeological and Genetic Perspectives on the Origins of Caste in South Asia," in *The Evolution and History of Human Populations in South Asia*, ed. Michael D. Petraglia and Bridget Allchin (Dordrecht, The Netherlands: Springer, 2007), 341–62.

36. Reich et al., "Reconstructing Indian Population History."

37. M. Arcos-Burgos and M. Muenke, "Genetics of Population Isolates," *Clinical Genetics* 61 (2002): 233–47.

38. N. Nakatsuka et al., "The Promise of Discovering Population-Specific Disease-Associated Genes in South Asia," *Nature Genetics* 49 (2017): 1403–7.

39. Reich et al., "Reconstructing Indian Population History."

40. I. Manoharan et al., "Naturally Occurring Mutation Leu307Pro of Human Butyrylcholinesterase in the Vysya Community of India," *Pharmacogenetics and Genomics* 16 (2006): 461–68.

41. A. E. Raz, "Can Population-Based Carrier Screening Be Left to the Community?," *Journal of Genetic Counseling* 18 (2009): 114–18.

42. I. Lazaridis et al., "Genomic Insights into the Origin of Farming in the Ancient Near East," *Nature* 536 (2016): 419–24; F. Broushaki et al., "Early Neolithic Genomes from the Eastern Fertile Crescent," *Science* 353 (2016): 499–503.

43. Ibid.

44. Lazaridis et al., "Genomic Insights."

45. Unpublished results from David Reich's laboratory.

7 In Search of Native American Ancestors

1. Betty Mindlin, *Unwritten Stories of the Suruí Indians of Rondônia* (Austin: Institute of Latin American Studies; distributed by the University of Texas Press, 1995).

2. D. Reich et al., "Reconstructing Native American Population History," *Nature* 488 (2012): 370–74.

3. P. Skoglund et al., "Genetic Evidence for Two Founding Populations of the Americas," *Nature* 525 (2015): 104–8.

4. P. D. Heintzman et al., "Bison Phylogeography Constrains Dispersal and Viability of the Ice Free Corridor in Western Canada," *Proceedings of the National Academy of Sciences of the U.S.A.* 113 (2016): 8057–63; M. W. Pedersen et al., "Postglacial Viability and Colonization in North America's Ice-Free Corridor," *Nature* 537 (2016): 45–49.

5. José de Acosta, *Historia Natural y Moral de las Indias: En que se Tratan las Cosas Notables del Cielo y Elementos, Metales, Plantas y Animales de Ellas y los Ritos, Ceremonias, Leyes y Gobierno y Guerras de los Indios* (Seville: Juan de León, 1590).

6. David J. Meltzer, *First Peoples in a New World: Colonizing Ice Age America* (Berkeley: University of California Press, 2009).

7. J. H. Greenberg, C. G. Turner II, and S. L. Zegura, "The Settlement of the Americas: A Comparison of the Linguistic, Dental, and Genetic Evidence," *Current Anthropology* 27 (1986): 477–97.

8. P. Forster, R. Harding, A. Torroni, and H.-J. Bandelt, "Origin and Evolution of Native American mtDNA Variation: A Reappraisal," *American Journal of Human Genetics* 59 (1996): 935–45; E. Tamm et al., "Beringian Standstill and Spread of Native American Founders," *PloS One* 2 (2017): e829.

9. T. D. Dillehay et al., "Monte Verde: Seaweed, Food, Medicine, and the Peopling of South America," *Science* 320 (2008): 784–86.

10. D. L. Jenkins et al., "Clovis Age Western Stemmed Projectile Points and Human Coprolites at the Paisley Caves," *Science* 337 (2012): 223–28.

11. M. Rasmussen et al., "The Genome of a Late Pleistocene Human from a Clovis Burial Site in Western Montana," *Nature* 506 (2014): 225–29.

12. Povos Indígenas No Brasil, "Karitiana: Biopiracy and the Unauthorized Collection of Biomedical Samples," https://pib.socioambiental.org/en/povo/karitiana/389.

13. N. A. Garrison and M. K. Cho, "Awareness and Acceptable Practices: IRB and Researcher Reflections on the Havasupai Lawsuit," *AJOB Primary Research* 4 (2013): 55–63; A. Harmon, "Indian Tribe Wins Fight to Limit Research of Its DNA," *New York Times*, April 21, 2010.

14. Ronald P. Maldonado, "Key Points for University Researchers When Considering a Research Project with the Navajo Nation," http://nptao.arizona.edu/sites/nptao/files/navajonationkeyresearchrequirements_0.pdf.

15. Rebecca Skloot, *The Immortal Life of Henrietta Lacks* (New York: Crown, 2010).

16. B. L. Shelton, "Consent and Consultation in Genetic Research on American Indians and Alaska Natives," http://www.ipcb.org/publications/briefing_papers/files/consent.html.

17. R. R. Sharp and M. W. Foster, "Involving Study Populations in the Review of Genetic Research," *Journal of Law, Medicine and Ethics* 28 (2000): 41–51; International HapMap Consortium, "The International HapMap Project," *Nature* 426 (2003): 789–96.

18. T. Egan, "Tribe Stops Study of Bones That Challenge History," *New York Times*, September 30, 1996; Douglas W. Owsley and Richard L. Jantz, *Kennewick Man: The Scientific Investigation of an Ancient American Skeleton* (College Station: Texas A&M University Press, 2014); D. J. Meltzer, "Kennewick Man: Coming to Closure," *Antiquity* 348 (2015): 1485–93.

19. M. Rasmussen et al., "The Ancestry and Affiliations of Kennewick Man," *Nature* 523 (2015): 455–58.

20. Ibid.

21. J. Lindo et al., "Ancient Individuals from the North American Northwest Coast Reveal 10,000 Years of Regional Genetic Continuity," *Proceedings of the National Academy of Sciences of the U.S.A.* 114 (2017): 4093–98.

22. Samuel J. Redman, *Bone Rooms: From Scientific Racism to Human Prehistory in Museums* (Cambridge, MA, and London: Harvard University Press, 2016).

23. M. Rasmussen et al., "An Aboriginal Australian Genome Reveals Separate Human Dispersals into Asia," *Science* 334 (2011): 94–98.

24. Rasmussen et al., "Genome of a Late Pleistocene Human."

25. Rasmussen et al., "Ancestry and Affiliations of Kennewick Man."

26. A. S. Malaspinas et al., "A Genomic History of Aboriginal Australia," *Nature* 538 (2016): 207–14.

27. E. Callaway, "Ancient Genome Delivers 'Spirit Cave Mummy' to US tribe," *Nature* 540 (2016): 178–79.

28. Ibid.

29. M. Livi-Bacci, "The Depopulation of Hispanic America After the Conquest," *Population and Development Review* 32 (2006): 199–232; Lewis H. Morgan, *Ancient Society; Or, Researches in the Lines of Human Progress from Savagery Through Barbarism to Civilization* (Chicago: Charles H. Kerr, 1909).

30. Reich et al., "Reconstructing Native American Population History."

31. Lindo et al., "Ancient Individuals."

32. Lyle Campbell and Marianne Mithun, *The Languages of Native America: Historical and Comparative Assessment* (Austin: University of Texas Press, 1979).

33. L. Campbell, "Comment on Greenberg, Turner and Zegura," *Current Anthropology* 27 (1986): 488.

34. Peter Bellwood, *First Migrants: Ancient Migration in Global Perspective* (Chichester, West Sussex, UK / Malden, MA: Wiley-Blackwell, 2013).

35. Reich et al., "Reconstructing Native American Population History."

36. W. A. Neves and M. Hubbe, "Cranial Morphology of Early Americans from Lagoa Santa, Brazil: Implications for the Settlement of the New World," *Proceedings of the National Academy of Sciences of the U.S.A.* 102 (2005): 18309–14.

37. Rasmussen et al., "Ancestry and Affiliations of Kennewick Man."

38. P. Skoglund et al., "Genetic Evidence for Two Founding Populations of the Americas," *Nature* 525 (2015): 104–8.

39. Povos Indígenas No Brasil, "Surui Paiter: Introduction," https://pib.socio ambiental.org/en/povo/surui-paiter; R. A. Butler, "Amazon Indians Use Google Earth, GPS to Protect Forest Home," *Mongabay: News and Inspiration from Nature's Frontline*, November 15, 2006, https://news.mongabay.com/2006/11/ amazon-indians-use-google-earth-gps-to-protect-forest-home/.

40. "Karitiana: Biopiracy and the Unauthorized Collection."

41. Povos Indígenas No Brasil, "Xavante: Introduction," https://pib.socioambiental .org/en/povo/xavante.

42. M. Raghavan et al., "Genomic Evidence for the Pleistocene and Recent Population History of Native Americans," *Science* 349 (2015): aab3884.

43. E. J. Vajda, "A Siberian Link with Na-Dene Languages," in *Anthropological Papers of the University of Alaska: New Series*, ed. James M. Kari and Ben Austin Potter, 5 (2010): 33–99.

44. Reich et al., "Reconstructing Native American Population History."

45. M. Rasmussen et al., "Ancient Human Genome Sequence of an Extinct Palaeo-Eskimo," *Nature* 463 (2010): 757–62.

46. M. Raghavan et al., "The Genetic Prehistory of the New World Arctic," *Science* 345 (2014): 1255832.

47. P. Flegontov et al., "Paleo-Eskimo Genetic Legacy Across North America," *bioRxiv* (2017): doi.org/10.1101.203018.

48. Flegontov et al., "Paleo-Eskimo Genetic Legacy."

49. T. M. Friesen, "Pan-Arctic Population Movements: The Early Paleo-Inuit and Thule Inuit Migrations," in *The Oxford Handbook of the Prehistoric Arctic*, ed. T. Max Friesen and Owen K. Mason (New York: Oxford University Press, 2016), 673–92.

50. Reich et al., "Reconstructing Native American Population History."

51. J. Diamond and P. Bellwood, "Farmers and Their Languages: The First Expansions," *Science* 300 (2003): 597–603; Peter Bellwood, *First Farmers: The Origins of Agricultural Societies* (Malden, MA: Blackwell, 2005).

52. R. R. da Fonseca et al., "The Origin and Evolution of Maize in the Southwestern United States," *Nature Plants* 1 (2015): 14003.

8 The Genomic Origins of East Asians

1. X. H. Wu et al., "Early Pottery at 20,000 Years Ago in Xianrendong Cave, China," *Science* 336 (2012): 1696–1700.

2. R. X. Zhu et al., "Early Evidence of the Genus *Homo* in East Asia," *Journal of Human Evolution* 55 (2008): 1075–85.

3. C. C. Swisher III et al., "Age of the Earliest Known Hominids in Java, Indonesia," *Science* 263 (1994): 1118–21; Peter Bellwood, *First Islanders: Prehistory and Human Migration in Island Southeast Asia* (Oxford: Wiley-Blackwell, 2017).

4. D. Richter et al., "The Age of the Hominin Fossils from Jebel Irhoud, Morocco, and the Origins of the Middle Stone Age," *Nature* 546 (2017): 293–96; J. G. Fleagle, Z. Assefa, F. H. Brown, and J. J. Shea, "Paleoanthropology of the Kibish Formation, Southern Ethiopia: Introduction," *Journal of Human Evolution* 55 (2008): 360–65.

5. T. Sutikna et al., "Revised Stratigraphy and Chronology for *Homo floresiensis* at Liang Bua in Indonesia," *Nature* 532 (2016): 366–69.

6. Y. Ke et al., "African Origin of Modern Humans in East Asia: A Tale of 12,000 Y Chromosomes," *Science* 292 (2001): 1151–53.

7. J. Qiu, "The Forgotten Continent: Fossil Finds in China Are Challenging Ideas About the Evolution of Modern Humans and Our Closest Relatives," *Nature* 535 (2016): 218–20.

8. R. J. Rabett and P. J. Piper, "The Emergence of Bone Technologies at the End of the Pleistocene in Southeast Asia: Regional and Evolutionary Implications," *Cambridge Archaeological Journal* 22 (2012): 37–56; M. C. Langley, C. Clarkson, and S. Ulm, "From Small Holes to Grand Narratives: The Impact of Taphonomy and Sample Size on the Modernity Debate in Australia and New Guinea," *Journal of Human Evolution* 61 (2011): 197–208; M. Aubert et al., "Pleistocene Cave Art from Sulawesi, Indonesia," *Nature* 514 (2014): 223–27.

9. Langley, Clarkson, and Ulm, "From Small Holes to Grand Narratives"; J. F. Connell and J. Allen, "The Process, Biotic Impact, and Global Implications of the Human Colonization of Sahul About 47,000 Years Ago," *Journal of Archaeological Science* 56 (2015): 73–84.

10. J.-J. Hublin, "The Modern Human Colonization of Western Eurasia: When and Where?," *Quaternary Science Reviews* 118 (2015): 194–210.

11. R. Foley and M. M. Lahr, "Mode 3 Technologies and the Evolution of Modern Humans," *Cambridge Archaeological Journal* 7 (1997): 3–36.

12. M. M. Lahr and R. Foley, "Multiple Dispersals and Modern Human Origins," *Evolutionary Anthropology* 3 (1994): 48–60.

13. H. Reyes-Centeno et al., "Testing Modern Human Out-of-Africa Dispersal Models and Implications for Modern Human Origins," *Journal of Human Evolution* 87 (2015): 95–106.

14. H. S. Groucutt et al., "Rethinking the Dispersal of *Homo sapiens* Out of Africa," *Evolutionary Anthropology* 24 (2015): 149–64.

15. R. Grün et al., "U-series and ESR Analyses of Bones and Teeth Relating to the Human Burials from Skhul," *Journal of Human Evolution* 49 (2005): 316–34.

16. S. J. Armitage et al., "The Southern Route 'Out of Africa': Evidence for an Early Expansion of Modern Humans into Arabia," *Science* 331 (2011): 453–56; M. D. Petraglia, "Trailblazers Across Africa," *Nature* 470 (2011): 50–51.

17. M. Kuhlwilm et al., "Ancient Gene Flow from Early Modern Humans into Eastern Neanderthals," *Nature* 530 (2016): 429–33.

18. M. Rasmussen et al., "An Aboriginal Australian Genome Reveals Separate Human Dispersals into Asia," *Science* 334 (2011): 94–98.

19. D. Reich et al., "Genetic History of an Archaic Hominin Group from Denisova Cave in Siberia," *Nature* 468 (2010): 1053–60; M. Meyer et al., "A High-Coverage Genome Sequence from an Archaic Denisovan Individual," *Science* 338 (2012): 222–26.

20. S. Mallick et al., "The Simons Genome Diversity Project: 300 Genomes from 142 Diverse Populations," *Nature* 538 (2016): 201–6.

21. Q. Fu et al., "Genome Sequence of a 45,000-Year-Old Modern Human from Western Siberia," *Nature* 514 (2014): 445–49; S. Sankararaman, S. Mallick, N. Patterson, and D. Reich, "The Combined Landscape of Denisovan and Neanderthal Ancestry in Present-Day Humans," *Current Biology* 26 (2016): 1241–47; P. Moorjani et al., " A Genetic Method for Dating Ancient Genomes Provides a Direct Estimate of Human Generation Interval in the Last 45,000 Years," *Proceedings of the National Academy of Sciences of the U.S.A.* 113 (2016): 5652–7.

22. Mallick et al., "Simons Genome Diversity Project"; M. Lipson and D. Reich, "A Working Model of the Deep Relationships of Diverse Modern Human Genetic Lineages Outside of Africa," *Molecular Biology and Evolution* 34 (2017): 889–902.

23. Mallick et al., "Simons Genome Diversity Project"; A. S. Malaspinas et al., "A Genomic History of Aboriginal Australia," *Nature* 538 (2016): 207–14; L. Pagani et al., "Genomic Analyses Inform on Migration Events During the Peopling of Eurasia," *Nature* 538 (2016): 238–42.

24. Hublin, "Modern Human Colonization of Western Eurasia."

25. M. Raghavan et al., "Upper Palaeolithic Siberian Genome Reveals Dual Ancestry of Native Americans," *Nature* (2013): doi: 10.1038/nature12736.

26. Hugo Pan-Asian SNP Consortium, "Mapping Human Genetic Diversity in Asia," *Science* 326 (2009): 1541–45.

27. S. Ramachandran et al., "Support from the Relationship of Genetic and Geographic Distance in Human Populations for a Serial Founder Effect Originating in Africa," *Proceedings of the National Academy of Sciences of the U.S.A.* 102 (2005): 15942–47; B. M. Henn, L. L. Cavalli-Sforza, and M. W. Feldman, "The Great Human Expansion," *Proceedings of the National Academy of Sciences of the U.S.A.* 109 (2012): 17758–64.

28. J. K. Pickrell and D. Reich, "Toward a New History and Geography of Human Genes Informed by Ancient DNA," *Trends in Genetics* 30 (2014): 377–89.

29. Unpublished results from David Reich's laboratory.

30. V. Siska et al., "Genome-Wide Data from Two Early Neolithic East Asian Individuals Dating to 7700 Years Ago," *Science Advances* 3 (2017): e1601877.

31. Peter Bellwood, *First Farmers: The Origins of Agricultural Societies* (Malden, MA: Blackwell, 2005).

32. J. Diamond and P. Bellwood, "Farmers and Their Languages: The First Expansions," *Science* 300 (2003): 597–603.

33. S. Xu et al., "Genomic Dissection of Population Substructure of Han Chinese and Its Implication in Association Studies," *American Journal of Human Genetics* 85 (2009): 762–74; J. M. Chen et al., "Genetic Structure of the Han Chinese Population Revealed by Genome-Wide SNP Variation," *American Journal of Human Genetics* 85 (2009): 775–85.

34. B. Wen et al., "Genetic Evidence Supports Demic Diffusion of Han Culture," *Nature* 431 (2004): 302–5.

35. F. H. Chen et al., "Agriculture Facilitated Permanent Human Occupation of the Tibetan Plateau After 3600 B.P.," *Science* 347 (2015): 248–50.

36. Unpublished results from David Reich's laboratory.

37. T. A. Jinam et al., "Unique Characteristics of the Ainu Population in Northern Japan," *Journal of Human Genetics* 60 (2015): 565–71.

38. Ibid.; P. R. Loh et al., "Inferring Admixture Histories of Human Populations Using Linkage Disequilibrium," *Genetics* 193 (2013): 1233–54.

39. Unpublished results from David Reich's laboratory; Bellwood, *First Migrants*.

40. Diamond and Bellwood, "Farmers and Their Languages."

41. M. Lipson et al., "Reconstructing Austronesian Population History in Island Southeast Asia," *Nature Communications* 5 (2014): 4689.

42. R. Blench, "Was There an Austroasiatic Presence in Island Southeast Asia Prior to the Austronesian Expansion?," *Bulletin of the Indo-Pacific Prehistory Association* 30 (2010): 133–44.

43. Bellwood, *First Migrants*.

44. A. Crowther et al., "Ancient Crops Provide First Archaeological Signature of the Westward Austronesian Expansion," *Proceedings of the National Academy of Sciences of the U.S.A.* 113 (2016): 6635–40.

45. Lipson et al., "Reconstructing Austronesian Population History."

46. A. Wollstein et al., "Demographic History of Oceania Inferred from Genome-Wide Data," *Current Biology* 20 (2010): 1983–92; M. Kayser, "The Human Genetic History of Oceania: Near and Remote Views of Dispersal," *Current Biology* 20 (2010): R194–201; E. Matisoo-Smith, "Ancient DNA and the Human Settlement of the Pacific: A Review," *Journal of Human Evolution* 79 (2015): 93–104.

47. D. Reich et al., "Denisova Admixture and the First Modern Human Dispersals into Southeast Asia and Oceania," *American Journal of Human Genetics* 89 (2011): 516–28; P. Skoglund et al., "Genomic Insights into the Peopling of the Southwest Pacific," *Nature* 538 (2016): 510–13.

48. R. Pinhasi et al., "Optimal Ancient DNA Yields from the Inner Ear Part of the Human Petrous Bone," *PLoS One* 10 (2015): e0129102.

49. Skoglund et al., "Genomic Insights."

50. Ibid.

51. Unpublished results from David Reich's laboratory and Johannes Krause's laboratory.

52. Ibid.

9 Rejoining Africa to the Human Story

1. J. Lachance et al., "Evolutionary History and Adaptation from High-Coverage Whole-Genome Sequences of Diverse African Hunter-Gatherers," *Cell* 150 (2012): 457–69.
2. V. Plagnol and J. D. Wall, "Possible Ancestral Structure in Human Populations," *PLoS Genetics* 2 (2006): e105; J. D. Wall, K. E. Lohmueller, and V. Plagnol, "Detecting Ancient Admixture and Estimating Demographic Parameters in Multiple Human Populations," *Molecular Biology and Evolution* 26 (2009): 1823–27.
3. M. F. Hammer et al., "Genetic Evidence for Archaic Admixture in Africa," *Proceedings of the National Academy of Sciences of the U.S.A.* 108 (2011): 15123–28.
4. K. Harvati et al., "The Later Stone Age Calvaria from Iwo Eleru, Nigeria: Morphology and Chronology," *PLoS One* 6 (2011): e24024; I. Crevecoeur, A. Brooks, I. Ribot, E. Cornelissen, and P. Semal, "The Late Stone Age Human Remains from Ishango (Democratic Republic of Congo): New Insights on Late Pleistocene Modern Human Diversity in Africa," *American Journal of Physical Anthropology* 96 (2016): 35–57.
5. Unpublished results from David Reich's laboratory.
6. D. Richter et al., "The Age of the Hominin Fossils from Jebel Irhoud, Morocco, and the Origins of the Middle Stone Age," *Nature* 546 (2017): 293–96; J. G. Fleagle, Z. Assefa, F. H. Brown, and J. J. Shea, "Paleoanthropology of the Kibish Formation, Southern Ethiopia: Introduction," *Journal of Human Evolution* 55 (2008): 360–65.
7. H. Li and R. Durbin, "Inference of Human Population History from Individual Whole-Genome Sequences," *Nature* 475 (2011): 493–96.
8. Li and Durbin, "Inference of Human Population History"; K. Prüfer et al., "The Complete Genome Sequence of a Neanderthal from the Altai Mountains," *Nature* (2013): doi: 10.1038/nature12886.
9. P. H. Dirks et al., "The Age of Homo Naledi and Associated Sediments in the Rising Star Cave, South Africa," *eLife* 6 (2017): e24231.
10. I. Gronau et al., "Bayesian Inference of Ancient Human Demography from Individual Genome Sequences," *Nature Genetics* 43 (2011): 1031–34.
11. P. Skoglund et al., "Reconstructing Prehistoric African Population Structure," *Cell* 171 (2017): 5694.
12. S. Mallick et al., "The Simons Genome Diversity Project: 300 Genomes from 142 Diverse Populations," *Nature* 538 (2016): 201–6; Gronau et al., "Bayesian Inference."
13. S. A. Tishkoff et al., "The Genetic Structure and History of Africans and African Americans," *Science* 324 (2009): 1035–44.
14. C. J. Holden, "Bantu Language Trees Reflect the Spread of Farming Across Sub-Saharan Africa: A Maximum-Parsimony Analysis," *Proceedings of the Royal Society B—Biological Sciences* 269 (2002): 793–99; P. de Maret, "Archaeologies of the Bantu Expansion," in *The Oxford Handbook of African Archaeology*, ed. Peter Mitchell and Paul J. Lane (Oxford: Oxford University Press, 2013), 627–44.
15. K. Bostoen et al., "Middle to Late Holocene Paleoclimatic Change and the Early Bantu Expansion in the Rain Forests of Western Central Africa," *Current*

Anthropology 56 (2016): 354–84; K. Manning et al., "4,500-Year-Old Domesticated Pearl Millet (*Pennisetum glaucum*) from the Tilemsi Valley, Mali: New Insights into an Alternative Cereal Domestication Pathway," *Journal of Archaeological Science* 38 (2011): 312–22.

16. D. Killick, "Cairo to Cape: The Spread of Metallurgy Through Eastern and Southern Africa," *Journal of World Prehistory* 22 (2009): 399–414.

17. de Maret, "Archaeologies of the Bantu Expansion."

18. Holden, "Bantu Language Trees."

19. Bostoen et al., "Middle to Late Holocene"; Manning et al., "4,500-Year-Old."

20. D. J. Lawson, G. Hellenthal, S. Myers, and D. Falush, "Inference of Population Structure Using Dense Haplotype Data," *PLoS Genetics* 8 (2012): e1002453; G. Hellenthal et al., "A Genetic Atlas of Human Admixture History," *Science* 343 (2014): 747–51; C. de Filippo, K. Bostoen, M. Stoneking, and B. Pakendorf, "Bringing Together Linguistic and Genetic Evidence to Test the Bantu Expansion," *Proceedings of the Royal Society B—Biological Sciences* 279 (2012): 3256–63; E. Patin et al., "Dispersals and Genetic Adaptation of Bantu-Speaking Populations in Africa and North America," *Science* 356 (2017): 543–46; G. B. Busby et al., "Admixture Into and Within Sub-Saharan Africa," *eLife* 5(2016): e15266.

21. Tishkoff et al., "Genetic Structure and History"; G. Ayodo et al., "Combining Evidence of Natural Selection with Association Analysis Increases Power to Detect Malaria-Resistance Variants," *American Journal of Human Genetics* 81 (2007): 234–42.

22. C. Ehret, "Reconstructing Ancient Kinship in Africa," in *Early Human Kinship: From Sex to Social Reproduction*, ed. Nicholas J. Allen, Hilary Callan, Robin Dunbar, and Wendy James (Malden, MA: Blackwell, 2008), 200–31; C. Ehret, S. O. Y. Keita, and P. Newman, "The Origins of Afroasiatic," *Science* 306 (2004): 1680–81.

23. J. Diamond and P. Bellwood, "Farmers and Their Languages: The First Expansions," *Science* 300 (2003): 597–603; P. Bellwood, "Response to Ehret et al. 'The Origins of Afroasiatic,'" *Science* 306 (2004): 1681.

24. D. Q. Fuller and E. Hildebrand, "Domesticating Plants in Africa," in *The Oxford Handbook of African Archaeology*, ed. Peter Mitchell and Paul J. Lane (Oxford: Oxford University Press, 2013), 507–26; M. Madella et al., "Microbotanical Evidence of Domestic Cereals in Africa 7000 Years Ago," *PLoS One* 9 (2014): e110177.

25. I. Lazaridis et al., "Genomic Insights into the Origin of Farming in the Ancient Near East," *Nature* 536 (2016): 419–24; Skoglund et al., "Reconstructing Prehistoric African Population Structure."

26. Lazaridis et al., "Genomic Insights"; Skoglund et al., "Reconstructing Prehistoric African Population Structure"; V. J. Schuenemann et al., "Ancient Egyptian Mummy Genomes Suggest an Increase of Sub-Saharan African Ancestry in Post-Roman Periods," *Nature Communications* 8 (2017): 15694.

27. T. Güldemann, "A Linguist's View: Khoe-Kwadi Speakers as the Earliest Food-Producers of Southern Africa," *Southern African Humanities* 20 (2008): 93–132.

28. J. K. Pickrell et al., "Ancient West Eurasian Ancestry in Southern and Eastern Africa," *Proceedings of the National Academy of Sciences of the U.S.A.* 111 (2014): 2632–37.

29. Pagani et al., "Ethiopian Genetic Diversity."

30. Skoglund et al., "Reconstructing Prehistoric African Population Structure."

31. Luigi Luca Cavalli-Sforza and Francesco Cavalli-Sforza, *The Great Human Diasporas: The History of Diversity and Evolution* (Reading, MA: Addison-Wesley, 1995).

32. M. Gallego Llorente et al., "Ancient Ethiopian Genome Reveals Extensive Eurasian Admixture Throughout the African Continent," *Science* 350 (2015): 820–22.

33. Donald N. Levine, *Greater Ethiopia: The Evolution of a Multiethnic Society* (Chicago: University of Chicago Press, 2000).

34. L. Van Dorp et al., "Evidence for a Common Origin of Blacksmiths and Cultivators in the Ethiopian Ari Within the Last 4500 Years: Lessons for Clustering-Based Inference," *PLoS Genetics* 11 (2015): e1005397.

35. D. Reich et al., "Reconstructing Indian Population History," *Nature* 461 (2009): 489–94.

36. Skoglund et al., "Reconstructing Prehistoric African Population Structure."

37. Ibid.

38. Ibid.

39. J. K. Pickrell et al., "The Genetic Prehistory of Southern Africa," *Nature Communications* 3 (2012): 1143; C. M. Schlebusch et al., "Genomic Variation in Seven Khoe-San Groups Reveals Adaptation and Complex African History," *Science* 338 (2012): 374–79; Mallick et al., "Simons Genome Diversity Project."

40. M. E. Prendergast et al., "Continental Island Formation and the Archaeology of Defaunation on Zanzibar, Eastern Africa," *PLoS One* 11 (2016): e0149565.

41. Skoglund et al., "Reconstructing Prehistoric African Population Structure."

42. P. Ralph and G. Coop, "Parallel Adaptation: One or Many Waves of Advance of an Advantageous Allele?," *Genetics* 186 (2010): 647–68.

43. S. A. Tishkoff et al., "Convergent Adaptation of Human Lactase Persistence in Africa and Europe," *Nature Genetics* 39 (2007): 31–40.

44. Ralph and Coop, "Parallel Adaptation."

10 The Genomics of Inequality

1. Peter Wade, *Race and Ethnicity in Latin America* (London and New York: Pluto Press, 2010).

2. Trans-Atlantic Slave Trade Database, www.slavevoyages.org/assessment/estimates.

3. K. Bryc et al., "The Genetic Ancestry of African Americans, Latinos, and European Americans Across the United States," *American Journal of Human Genetics* 96 (2015): 37–53.

4. Piers Anthony, *Race Against Time* (New York: Hawthorn Books, 1973).

5. The first federal census in 1790 recorded 292,627 male slaves in Virginia out of a total male population of 747,610; available online at www.nationalgeographic.org/media/us-census-1790/.

6. Joshua D. Rothman, *Notorious in the Neighborhood: Sex and Families Across the Color Line in Virginia, 1787–1861* (Chapel Hill: University of North Carolina Press, 2003).

7. E. A. Foster et al., "Jefferson Fathered Slave's Last Child," *Nature* 396 (1998): 27–28.

8. "Statement on the TJMF Research Committee Report on Thomas Jeffer-

son and Sally Hemings," January 26, 2000, available online at https://www.monticello.org/sites/default/files/inline-pdfs/jefferson-hemings_report.pdf.

9. M. Hemings, "Life Among the Lowly, No. 1," *Pike County (Ohio) Republican*, March 13, 1873.

10. E. J. Parra et al., "Ancestral Proportions and Admixture Dynamics in Geographically Defined African Americans Living in South Carolina," *American Journal of Physical Anthropology* 114 (2001): 18–29.

11. Ibid.

12. Bryc et al., "Genetic Ancestry."

13. J. N. Fenner, "Cross-Cultural Estimation of the Human Generation Interval for Use in Genetics-Based Population Divergence Studies," *American Journal of Physical Anthropology* 128 (2005): 415–23.

14. David Morgan, *The Mongols* (Malden, MA, and Oxford: Blackwell, 2007).

15. T. Zerjal et al., "The Genetic Legacy of the Mongols," *American Journal of Human Genetics* 72 (2003): 717–21.

16. L. T. Moore et al., "A Y-Chromosome Signature of Hegemony in Gaelic Ireland," *American Journal of Human Genetics* 78 (2006): 334–38.

17. S. Lippold et al., "Human Paternal and Maternal Demographic Histories: Insights from High-Resolution Y Chromosome and mtDNA Sequences," *Investigative Genetics* 5 (2014): 13; M. Karmin et al., "A Recent Bottleneck of Y Chromosome Diversity Coincides with a Global Change in Culture," *Genome Research* 25 (2015): 459–66.

18. Ibid.

19. A. Sherratt, "Plough and Pastoralism: Aspects of the Secondary Products Revolution," in *Pattern of the Past: Studies in Honour of David Clarke*, ed. Ian Hodder, Glynn Isaac, and Norman Hammond (Cambridge: Cambridge University Press, 1981), 261–306.

20. David W. Anthony, *The Horse, the Wheel, and Language: How Bronze-Age Riders from the Eurasian Steppes Shaped the Modern World* (Princeton, NJ: Princeton University Press, 2007).

21. W. Haak et al., "Massive Migration from the Steppe Was a Source for Indo-European Languages in Europe," *Nature* 522 (2015): 207–11; M. E. Allentoft et al., "Population Genomics of Bronze Age Eurasia," *Nature* 522 (2015): 167–72.

22. E. Murphy and A. Khokhlov, "A Bioarchaeological Study of Prehistoric Populations from the Volga Region," in *A Bronze Age Landscape in the Russian Steppes: The Samara Valley Project, Monumenta Archaeologica 37*, ed. David W. Anthony, Dorcas R. Brown, Aleksandr A. Khokhlov, Pavel V. Kuznetsov, and Oleg D. Mochalov (Los Angeles: Cotsen Institute of Archaeology Press, 2016), 149–216.

23. Marija Gimbutas, *The Prehistory of Eastern Europe, Part I: Mesolithic, Neolithic and Copper Age Cultures in Russia and the Baltic Area* (American School of Prehistoric Research, Harvard University, Bulletin No. 20) (Cambridge, MA: Peabody Museum, 1956).

24. Haak et al., "Massive Migration."

25. R. S. Wells et al., "The Eurasian Heartland: A Continental Perspective on Y-Chromosome Diversity," *Proceedings of the National Academy of Sciences of the U.S.A.* 98 (2001): 10244–49.

26. R. Martiniano et al., "The Population Genomics of Archaeological Transition in West Iberia: Investigation of Ancient Substructure Using Imputation and Haplotype-Based Methods," *PLoS Genetics* 13 (2017): e1006852.

27. M. Silva et al., "A Genetic Chronology for the Indian Subcontinent Points to Heavily Sex-Biased Dispersals," *BMC Evolutionary Biology* 17 (2017): 88.

28. Martiniano et al., "West Iberia"; unpublished results from David Reich's laboratory.

29. J. A. Tennessen et al., "Evolution and Functional Impact of Rare Coding Variation from Deep Sequencing of Human Exomes," *Science* 337 (2012): 64–69.

30. A. Keinan, J. C. Mullikin, N. Patterson, and D. Reich, "Accelerated Genetic Drift on Chromosome X During the Human Dispersal out of Africa," *Nature Genetics* 41 (2009): 66–70; A. Keinan and D. Reich, "Can a Sex-Biased Human Demography Account for the Reduced Effective Population Size of Chromosome X in Non-Africans?," *Molecular Biology and Evolution* 27 (2010): 2312–21.

31. P. Verdu et al., "Sociocultural Behavior, Sex-Biased Admixture, and Effective Population Sizes in Central African Pygmies and Non-Pygmies," *Molecular Biology and Evolution* 30 (2013): 918–37.

32. S. Mallick et al., "The Simons Genome Diversity Project: 300 Genomes from 142 Diverse Populations," *Nature* 538 (2016): 201–6.

33. L. G. Carvajal -Carmona et al., "Strong Amerind/White Sex Bias and a Possible Sephardic Contribution Among the Founders of a Population in Northwest Colombia," *American Journal of Human Genetics* 67 (2000): 1287–95.

34. Bedoya et al., "Admixture Dynamics in Hispanics: A Shift in the Nuclear Genetic Ancestry of a South American Population Isolate," *Proceedings of the National Academy of Sciences of the U.S.A.* 103 (2006): 7234–39.

35. P. Moorjani et al., "Genetic Evidence for Recent Population Mixture in India," *American Journal of Human Genetics* 93 (2013): 422–38.

36. M. Bamshad et al., "Genetic Evidence on the Origins of Indian Caste Populations," *Genome Research* 11 (2001): 994–1004; D. Reich et al., "Reconstructing Indian Population History," *Nature* 461 (2009): 489–94.

37. Bamshad et al., "Genetic Evidence"; I. Thanseem et al., "Genetic Affinities Among the Lower Castes and Tribal Groups of India: Inference from Y Chromosome and Mitochondrial DNA," *BMC Genetics* 7 (2006): 42.

38. M. Kayser, "The Human Genetic History of Oceania: Near and Remote Views of Dispersal," *Current Biology* 20 (2010): R194–201; P. Skoglund et al., "Genomic Insights into the Peopling of the Southwest Pacific," *Nature* 538 (2016): 510–13.

39. F. M. Jordan, R. D. Gray, S. J. Greenhill, and R. Mace, "Matrilocal Residence Is Ancestral in Austronesian Societies," *Proceedings of the Royal Society B—Biological Sciences* 276 (2009): 1957–64.

40. Skoglund et al., "Genomic Insights."

41. I. Lazaridis and D. Reich, "Failure to Replicate a Genetic Signal for Sex Bias in the Steppe Migration into Central Europe," *Proceedings of the National Academy of Sciences of the U.S.A.* 114 (2017): E3873–74.

11 The Genomics of Race and Identity

1. Centers for Disease Control and Prevention, "Prostate Cancer Rates by Race and Ethnicity," https://www.cdc.gov/cancer/prostate/statistics/race.htm.

2. N. Patterson et al., "Methods for High-Density Admixture Mapping of Disease

Genes," *American Journal of Human Genetics* 74 (2004): 979–1000; M. W. Smith et al., "A High-Density Admixture Map for Disease Gene Discovery in African Americans," *American Journal of Human Genetics* 74 (2004): 1001–13.

3. M. L. Freedman et al., "Admixture Mapping Identifies 8q24 as a Prostate Cancer Risk Locus in African-American Men," *Proceedings of the National Academy of Sciences of the U.S.A.* 103 (2006): 14068–73.

4. C. A. Haiman et al., "Multiple Regions within 8q24 Independently Affect Risk for Prostate Cancer," *Nature Genetics* 39 (2007): 638–44.

5. Freedman et al., "Admixture Mapping Identifies 8q24."

6. M. F. Ashley Montagu, *Man's Most Dangerous Myth: The Fallacy of Race* (New York: Columbia University Press, 1942).

7. R. C. Lewontin, "The Apportionment of Human Diversity," *Evolutionary Biology* 6 (1972): 381–98.

8. J. M. Stevens, "The Feasibility of Government Oversight for NIH-Funded Population Genetics Research," in *Revisiting Race in a Genomic Age* (Studies in Medical Anthropology), ed. Barbara A. Koenig, Sandra Soo-Jin Lee, and Sarah S. Richardson (New Brunswick, NJ: Rutgers University Press, 2008), 320–41; J. Stevens, "Racial Meanings and Scientific Methods: Policy Changes for NIH-Sponsored Publications Reporting Human Variation," *Journal of Health Policy, Politics and Law* 28 (2003): 1033–87.

9. N. A. Rosenberg et al., "Genetic Structure of Human Populations," *Science* 298 (2002): 2381–85.

10. D. Serre and S. Pääbo, "Evidence for Gradients of Human Genetic Diversity Within and Among Continents," *Genome Research* 14 (2004): 1679–85; F. B. Livingstone, "On the Non-Existence of Human Races," *Current Anthropology* 3 (1962): 279.

11. J. Dreyfuss, "Getting Closer to Our African Origins," *The Root*, October 17, 2011, www.theroot.com/getting-closer-to-our-african-origins-1790866394.

12. N. A. Rosenberg et al., "Clines, Clusters, and the Effect of Study Design on the Inference of Human Population Structure," *PLoS Genetics* 1 (2005): e70.

13. E. G. Burchard et al., "The Importance of Race and Ethnic Background in Biomedical Research and Clinical Practice," *New England Journal of Medicine* 348 (2003): 1170–75.

14. J. F. Wilson et al., "Population Genetic Structure of Variable Drug Response," *Nature Genetics* 29 (2001): 265–69.

15. D. Fullwiley, "The Biologistical Construction of Race: 'Admixture' Technology and the New Genetic Medicine," *Social Studies of Science* 38 (2008): 695–735.

16. Lewontin, "The Apportionment of Human Diversity"; A. R. Templeton, "Biological Races in Humans," *Studies in History and Philosophy of Biological and Biomedical Science* 44 (2013): 262–71.

17. *Razib Khan*, www.razib.com/wordpress.

18. *Dienekes' Anthropology Blog*, dienekes.blogspot.com.

19. *Eurogenes Blog*, http://eurogenes.blogspot.com.

20. Léon Poliakov, *The Aryan Myth: A History of Racist and Nationalist Ideas in Europe* (New York: Basic Books, 1974).

21. B. Arnold, "The Past as Propaganda: Totalitarian Archaeology in Nazi Germany," *Antiquity* 64 (1990): 464–78.

22. J. K. Pritchard, J. K. Pickrell, and G. Coop, "The Genetics of Human Adapta-

tion: Hard Sweeps, Soft Sweeps, and Polygenic Adaptation," *Current Biology* 20 (2010): R208–15; R. D. Hernandez et al., "Classic Selective Sweeps Were Rare in Recent Human Evolution," *Science* 331 (2011): 920–24.

23. M. C. Turchin et al., "Evidence of Widespread Selection on Standing Variation in Europe at Height-Associated SNPs," *Nature Genetics* 44 (2012): 1015–19.

24. Y. Field et al., "Detection of Human Adaptation During the Past 2000 Years," *Science* 354 (2016): 760–64.

25. A. Okbay et al., "Genome-Wide Association Study Identifies 74 Loci Associated with Educational Attainment," *Nature* 533 (2016): 539–42.

26. To compute the expected difference in number of years of education between the highest 5 percent and lowest 5 percent of genetically predicted educational attainment based on the numbers in the 2016 study by Benjamin and colleagues, I performed the following computation: (1) The number of years of education in the cohort analyzed by Benjamin and colleagues is quoted as 14.3 ± 3.7. I estimated the standard deviation of 3.7 years from the fact that the study estimates the effect size in weeks to be "0.014 to 0.048 standard deviations per allele (2.7 to 9.0 weeks of schooling)." These numbers translate to 188 (= 9.0 / 0.048) to 193 (= 2.7 / 0.014) weeks. Dividing by 52 weeks per year gives 3.7. (2) Benjamin and colleagues also report a genetic predictor of number of years of education that explains 3.2 percent of the variance of the trait. Therefore, the correlation between the predicted value and the actual value is $\sqrt{0.032} = 0.18$. We can model this mathematically using a two-dimensional normal distribution. (3) The probability that a person who is in the bottom 5% of the predicted distribution (more than 1.64 standard deviations below the average) has more than 12 years of education is then given by the proportion of people who are in the bottom 5 percent of the predicted distribution and also have more than 12 years of education (which can be calculated by measuring the area of the two-dimensional normal distribution that matches these criteria), divided by 0.05. This gives a probability of 60 percent. A similar calculation for the proportion of people in the top 5 percent of the predicted distribution gives a probability of 84 percent. (4) The Benjamin study also suggests that with enough samples it would be possible to build a reliable genetic predictor that accounts for 20 percent of the variance. Redoing the calculation using 20 percent instead of 3.2 percent leads to a prediction that 37 percent of people in the bottom 5 percent of the predicted distribution would complete twelve years of education compared to 96 percent of the top 5 percent.

27. A. Kong et al., "Selection Against Variants in the Genome Associated with Educational Attainment," *Proceedings of the National Academy of Sciences of the U.S.A.* 114 (2017): E727–32.

28. Kong et al., "Selection Against Variants," estimate that the genetically predicted number of years of education has decreased by an estimated 0.1 standard deviations over the last century under the pressure of natural selection.

29. G. Davies et al., "Genome-Wide Association Study of Cognitive Functions and Educational Attainment in UK Biobank (N=112 151)," *Molecular Psychiatry* 21 (2016): 758–67; M. T. Lo et al., "Genome-Wide Analyses for Personality Traits Identify Six Genomic Loci and Show Correlations with Psychiatric Disorders," *Nature Genetics* 49 (2017): 152–56.

30. S. Sniekers et al., "Genome-Wide Association Meta-Analysis of 78,308 Individ-

uals Identifies New Loci and Genes Influencing Human Intelligence," *Nature Genetics* 49 (2017): 1107–12.

31. I. Mathieson et al., "Genome-wide Patterns of Selection in 230 Ancient Eurasians," *Nature* 528 (2015): 499–503; Field et al., "Detection of Human Adaptation."

32. N. A. Rosenberg et al., "Genetic Structure of Human Populations," *Science* 298 (2002): 2381–85.

33. S. Ramachandran et al., "Support from the Relationship of Genetic and Geographic Distance in Human Populations for a Serial Founder Effect Originating in Africa," *Proceedings of the National Academy of Sciences of the U.S.A.* 102 (2005): 15942–47; B. M. Henn, L. L. Cavalli-Sforza, and M. W. Feldman, "The Great Human Expansion," *Proceedings of the National Academy of Sciences of the U.S.A.* 109 (2012): 17758–64.

34. J. K. Pickrell and D. Reich, "Toward a New History and Geography of Human Genes Informed by Ancient DNA," *Trends in Genetics* 30 (2014): 377–89.

35. M. Raghavan et al., "Upper Palaeolithic Siberian Genome Reveals Dual Ancestry of Native Americans," *Nature* (2013): doi: 10.1038/nature 12736.

36. I. Lazaridis et al., "Genomic Insights into the Origin of Farming in the Ancient Near East," *Nature* 536 (2016): 419–24.

37. Nicholas Wade, *A Troublesome Inheritance: Genes, Race and Human History* (New York: Penguin Press, 2014).

38. G. Coop et al., "A Troublesome Inheritance" (letters to the editor), *New York Times*, August 8, 2014.

39. G. Cochran, J. Hardy, and H. Harpending, "Natural History of Ashkenazi Intelligence," *Journal of Biosocial Science* 38 (2006): 659–93.

40. P. F. Palamara, T. Lencz, A. Darvasi, and I. Pe'er, "Length Distributions of Identity by Descent Reveal Fine-Scale Demographic History," *American Journal of Human Genetics* 91 (2012): 809–22; M. Slatkin, "A Population-Genetic Test of Founder Effects and Implications for Ashkenazi Jewish Diseases," *American Journal of Human Genetics* 75 (2004): 282–93.

41. H. Harpending, "The Biology of Families and the Future of Civilization" (minute 38), Preserving Western Civilization, 2009 Conference, audio available at www.preservingwesternciv.com/audio/07%20Prof._Henry_Harpending--The_Biology_of_Families_and_the_Future_of_Civilization.mp3 (2009).

42. G. Clark, "Genetically Capitalist? The Malthusian Era, Institutions and the Formation of Modern Preferences" (2007), www.econ.ucdavis.edu/faculty/gclark/papers/Capitalism%20Genes.pdf; Gregory Clark, *A Farewell to Alms: A Brief Economic History of the World* (Princeton, NJ: Princeton University Press, 2007).

43. Wade, *A Troublesome Inheritance*.

44. C. Hunt-Grubbe, "The Elementary DNA of Dr. Watson," *The Sunday Times*, October 14, 2017.

45. Coop et al. letters, *New York Times*.

46. David Epstein, *The Sports Gene: Inside the Science of Extraordinary Athletic Performance* (New York: Current, 2013).

47. Ibid.

48. I performed this computation as follows. (1) The 99.9999999th percentile of a trait corresponds to 6.0 standard deviations from the mean, whereas

the 99.99999th percentile corresponds to 5.2 standard deviations. Thus a 0.8-standard-deviation shift corresponds to a hundredfold enrichment of individuals. (2) I assumed that the 1.33-fold higher genetic variation in sub-Saharan Africans applies not just to random mutations in the genome, but also to mutations modulating biological traits. The standard deviation is thus expected to be $1.15 = \sqrt{1.33}$-fold higher in sub-Saharan Africans based on a formula in J. J. Berg and G. Coop, "A Population Genetic Signal of Polygenic Adaptation," *PLoS Genetics* 10 (2014): e1004412, so the 6.0-standard-deviation cutoff in non-Africans corresponds to $5.2 = 6.0 / 1.15$ of that in sub-Saharan Africans, leading to the same predicted hundredfold enrichment above the 99.9999999th percentile.

49. W. Haak et al., "Massive Migration from the Steppe Was a Source for Indo-European Languages in Europe," *Nature* 522 (2015): 207–11; M. E. Allentoft et al., "Population Genomics of Bronze Age Eurasia," *Nature* 522 (2015): 167–72.

50. D. Reich et al., "Reconstructing Indian Population History," *Nature* 461 (2009): 489–94; Lazaridis et al., "Genomic Insights."

51. Michael F. Robinson, *The Lost White Tribe: Explorers, Scientists, and the Theory That Changed a Continent* (New York: Oxford University Press, 2016).

52. Alex Haley, *Roots: The Saga of an American Family* (New York: Doubleday, 1976).

53. "Episode 4: (2010) Know Thyself" (minute 17) in *Faces of America with Henry Louis Gates Jr.*, http://www.pbs.org/wnet/facesofamerica/video/episode-4-know-thyself/237/.

54. African Ancestry, "Frequently Asked Questions," "About the Results," question 3 (2016), http://www.africanancestry.com/faq/.

55. Dreyfuss, "Getting Closer to Our African Origins."

56. S. Sailer, "African Ancestry Inc. Traces DNA Roots," United Press International, April 28, 2003, www.upi.com/inc/view.php?StoryID=20030428-074922-7714r.

57. Unpublished results from David Reich's laboratory.

58. H. Schroeder et al., "Genome-Wide Ancestry of 17th-Century Enslaved Africans from the Caribbean," *Proceedings of the National Academy of Sciences of the U.S.A.* 112 (2015): 3669–73.

59. R. E. Green et al., "A Draft Sequence of the Neanderthal Genome," *Science* 328 (2010): 710–22.

60. E. Durand, 23andMe: "White Paper 23-05: Neanderthal Ancestry Estimator" (2011), https://23andme.https.internapcdn.net/res/pdf/hXitekfSJe1lcIy7-Q72XA_23-05_Neanderthal_Ancestry.pdf; S. Sankararaman et al., "The Genomic Landscape of Neanderthal Ancestry in Present-Day Humans," *Nature* 507 (2014): 354–57.

61. Sankararaman et al., "Genomic Landscape."

62. https://customercare.23andme.com/hc/en-us/articles/212873707-Neanderthal-Report-Basics, #13514.

12   The Future of Ancient DNA

1. J. R. Arnold and W. F. Libby, "Age Determinations by Radiocarbon Content—Checks with Samples of Known Age," *Science* 110 (1949): 678–80.

2. Colin Renfrew, *Before Civilization: The Radiocarbon Revolution and Prehistoric Europe* (London: Jonathan Cape, 1973).

3. Lewis R. Binford, *In Pursuit of the Past: Decoding the Archaeological Record* (Berkeley: University of California Press, 1983).

4. M. Rasmussen et al., "Ancient Human Genome Sequence of an Extinct Palaeo-Eskimo," *Nature* 463 (2010): 757–62; M. Rasmussen et al., "The Genome of a Late Pleistocene Human from a Clovis Burial Site in Western Montana," *Nature* 506 (2014): 225–29; M. Raghavan et al., "Upper Palaeolithic Siberian Genome Reveals Dual Ancestry of Native Americans," *Nature* (2013): doi: 10.1038/nature 12736.

5. P. Skoglund et al., "Genomic Insights into the Peopling of the Southwest Pacific," *Nature* 538 (2016): 510–13.

6. J. Dabney et al., "Complete Mitochondrial Genome Sequence of a Middle Pleistocene Cave Bear Reconstructed from Ultrashort DNA Fragments," *Proceedings of the National Academy of Sciences of the U.S.A.* 110 (2013): 15758–63; M. Meyer et al., "A High-Coverage Genome Sequence from an Archaic Denisovan Individual," *Science* 338 (2012): 222–26; Q. Fu et al., "DNA Analysis of an Early Modern Human from Tianyuan Cave, China," *Proceedings of the National Academy of Sciences of the U.S.A.* 110 (2013): 2223–27; R. Pinhasi et al., "Optimal Ancient DNA Yields from the Inner Ear Part of the Human Petrous Bone," *PLoS One* 10 (2015): e0129102.

7. I. Lazaridis et al., "Genomic Insights into the Origin of Farming in the Ancient Near East," *Nature* 536 (2016): 419–24.

8. I. Olalde et al., "The Beaker Phenomenon and the Genomic Transformation of Northwest Europe," *bioRxiv* (2017): doi.org/10.1101/135962.

9. P. F. Palamara, T. Lencz, A. Darvasi, and I. Pe'er, "Length Distributions of Identity by Descent Reveal Fine-Scale Demographic History," *American Journal of Human Genetics* 91 (2012): 809–22; D. J. Lawson, G. Hellenthal, S. Myers, and D. Falush, "Inference of population structure using dense haplotype data," *PLoS Genetics* 8 (2012): e1002453.

10. S. Leslie et al., "The Fine-Scale Genetic Structure of the British Population," *Nature* 519 (2015): 309–14.

11. S. R. Browning and B. L. Browning, "Accurate Non-parametric Estimation of Recent Effective Population Size from Segments of Identity by Descent," *American Journal of Human Genetics* 97 (2015): 404–18.

12. M. Lynch, "Rate, Molecular Spectrum, and Consequences of Human Mutation," *Proceedings of the National Academy of Sciences of the U.S.A.* 107 (2010): 961–68; A. Kong et al., "Selection Against Variants in the Genome Associated with Educational Attainment," *Proceedings of the National Academy of Sciences of the U.S.A.* 114 (2017): E727–32.

13. J. K. Pritchard, J. K. Pickrell, and G. Coop, "The Genetics of Human Adaptation: Hard Sweeps, Soft Sweeps, and Polygenic Adaptation," *Current Biology* 20 (2010): R208–15.

14. S. Haensch et al., "Distinct Clones of *Yersinia pestis* Caused the Black Death," *PLoS Pathogens* 6 (2010): e1001134; K. I. Bos et al., "A Draft Genome of *Yersinia pestis* from Victims of the Black Death," *Nature* 478 (2011): 506–10.

15. I. Wiechmann and G. Grupe, "Detection of *Yersinia pestis* DNA in Two Early Medieval Skeletal Finds from Aschheim (Upper Bavaria, 6th Century AD),"

American Journal of Physical Anthropology 126 (2005): 48–55; D. M. Wagner et al., "*Yersinia pestis* and the Plague of Justinian 541–543 AD: A Genomic Analysis," *Lancet Infectious Diseases* 14 (2014): 319–26.

16. S. Rasmussen et al., "Early Divergent Strains of *Yersinia pestis* in Eurasia 5,000 Years Ago," *Cell* 163 (2015): 571–82.

17. P. Singh et al., "Insight into the Evolution and Origin of Leprosy Bacilli from the Genome Sequence of *Mycobacterium lepromatosis*," *Proceedings of the National Academy of Sciences of the U.S.A.* 112 (2015): 4459–64.

18. K. I. Bos et al., "Pre-Columbian Mycobacterial Genomes Reveal Seals as a Source of New World Human Tuberculosis," *Nature* 514 (2014): 494–97.

19. K. Yoshida et al., "The Rise and Fall of the *Phytophthora infestans* Lineage That Triggered the Irish Potato Famine," *eLife* 2 (2013): e00731.

20. C. Warinner et al., "Pathogens and Host Immunity in the Ancient Human Oral Cavity," *Nature Genetics* 46 (2014): 336–44.

21. T. Higham et al., "The Timing and Spatiotemporal Patterning of Neanderthal Disappearance," *Nature* 512 (2014): 306–9.

22. E. Callaway, "Ancient Genome Delivers 'Spirit Cave Mummy' to US Tribe," *Nature* 540 (2016): 178–79.

Index

Page numbers in *italics* refer to illustrations.